Adaptive Resource Management in Active Network Nodes

Yuhong Li

Vom Fachbereich für Mathematik und Informatik
der Technischen Universität Braunschweig
genehmigte Dissertation
zur Erlangung des Grades einer
Doktor-Ingenieurin (Dr.-Ing.)

Tag der mündlichen Prüfung: 18. Juni 2004

Erster Referent: Prof. Dr.-Ing. Lars C. Wolf

Zweiter Referent: Prof. Dr.-Ing. Jens B. Schmitt

Bibliografische Information Der Deutschen Bibliothek

Die Deutsche Bibliothek verzeichnet diese Publikation in der Deutschen
Nationalbibliografie; detaillierte bibliografische Daten sind im Internet über
http://dnb.ddb.de abrufbar.

ISBN 3-8325-0633-0

Logos Verlag Berlin
Comeniushof, Gubener Str. 47,
10243 Berlin
Tel.: +49 030 42 85 10 90
Fax: +49 030 42 85 10 92
INTERNET: http://www.logos-verlag.de

Gedruckt mit Unterstützung des Deutschen
Akademischen Austauschdienstes

Kurzfassung

Aktive Netze ermöglichen die Injektion und Ausführung von benutzerspezifischen Programmen innerhalb der Netzinfrastruktur um eine schnelle Fortentwicklung der Netzdienste und Protokolle zu ermöglichen und das derzeitige Internet flexibler zu machen. In bisherigen Forschungsansätzen sind Aspekte wie Design und Evaluierung von Programmausführungsumgebungen und der unterliegenden Betriebssysteme für aktive Netzknoten untersucht werden. Die Ressourcenverwaltung wurde dagegen nur im Zusammenhang mit Sicherheitsaspekten erforscht. Dazu zählen z.b. das Verhindern der Überlastung von Systemressourcen durch benutzerdefinierte Programme oder die Isolierung von Benutzerprogrammprozessen in aktiven Netzknoten. Die Bereitstellung von Ressourcegarantien für die Ausführung von benutzerdefinierten Programmen in aktiven Netzknoten wurde dagegen nicht betrachtet.

Durch die Einführung von benutzerspezifischen Berechnungen in aktiven Netzknoten können solche Ressourcen wie Prozessor und Speicher nicht mehr vernachlässigt werden, wie es in traditionellen Netzen der Fall ist. Vielmehr kann die Ressourcenutzung in aktiven Netzen durch die Belegung der unterschiedlichen Ressourcentypen von diversen Benutzerprogrammen stark unterschiedlich sein. Das kann dazu führen, dass oft Situationen auftreten, in denen benutzerdefinierte Berechnungen aufgrund des Fehlens eines bestimmten Ressourcentyps nicht ausgeführt werden können. Aus diesem Grund können die im derzeitigen Internet eingesetzten Methoden der Ressourcenverwaltung nicht auf aktive Netze übertragen werden. Stattdessen sind neue Methoden zur Ressourcenverwaltung in aktiven Netzen notwendig. Es sollten solche Aspekte wie die Nutzung einzelner Ressourcentypen, mögliche Kompromisse in der Auslastung von unterschiedlichen Ressourcen und die totale Systemressourcenauslastung berücksichtigt werden.

Im Rahmen dieser Arbeit werden neue Verfahren für die Ressourcenverwaltung in aktiven Netzen entwickelt. Es wird das Konzept eines Ressourcenvektors vorgestellt, das eine Beschreibung der Ressourcenausnutzung in aktiven Netzen ermöglicht, indem sowohl der Typ als auch die Anzahl der vorhandenen Ressourcen in einem aktiven Netzknoten erfasst werden. Darauf basierend wird der Vektorraum anpassbarer Ressourcen (ARVS) eingeführt, um die Ressourcen-adaptivität der Anwendungen zu beschreiben. Der ARVS formuliert ein generisches Modell für die Adaptation in Netzknoten und drückt die individuellen Ressourcenanforderungen der Anwendungen aus.

Des Weiteren wird ein adaptiver Mechanismus für die Zugangskontrolle entwickelt, der eine Unterstützung der Ressourcenadaption zwischen den einzelnen Ressourcentypen und Anwendungen in aktiven Netzknoten ermöglicht. Dieser neue Mechanismus berücksichtigt nicht nur die Performanz der Anwendungen sondern auch die Auslastung der verschiedenen Ressourcentypen in einem aktiven Netzknoten. Es werden sowohl die generische Netzzugangskontrolle als auch flexible anwendungsspezifische Adaptationen in Netzknoten unterstützt.

Darüber hinaus wird eine Architektur eines aktiven Netzknotens mit Unterstützung des vorgeschlagenen Mechanismus für die adaptive Zugangskontrolle implementiert. Das Ressourcenverwaltungssubsystem innerhalb der Architektur beinhaltet den Algorithmus zur Zugangskontrolle und führt die Ressourcenadaption durch. Es überwacht und steuert die gesamte Ressourcennutzung eines aktiven Netzknotens. Basierend auf der Architektur werden einige experimentelle Anwendungen vorgestellt und Resultate der damit durchgeführten Performanzevaluierung des implementierten Netwerkknotens präsentiert.

Acknowledgment

I would like to express my deepest gratitude towards my supervisor, Prof. Lars Wolf, for his boundless patience and effective advice throughout this research work. I appreciate his insightful suggestions and comments to the whole work and the careful reading and corrections done to this dissertation despite his busy schedules. This dissertation would not have been possible without his guidance and support. Prof. Lars Wolf had also shown me many soft skills possessed by a capable researcher. His enthusiasm for work and working methods will influence me also in the future. I am also deeply indebted to him for supporting me to attend academic conferences and helping me in all aspects during my stay in Germany.

Prof. Jens Schmitt deserves particular thanks for his review of the dissertation and many valuable comments.

I would like also to thank Prof. Ralf Steinmetz for his enormous help at the beginning of my study in Germany. I am thankful to my colleagues at the Institute of Operating Systems and Computer Networks, Technical University of Braunschweig, and the Institute of Telematics, University of Karlsruhe, who organized different activities for technical exchange. Particularly, I want to thank Marc Bechler for his all kinds of helps during my time in Karlsruhe and Braunschweig, and also Zefir Kurtisi, Andreas Kleinschmidt, Matthias Dick for reading the drafts of this dissertation. My thanks go to Frank Strauß, Verena Kahmann and Ulrich Timm for their help during my preparation for the tests. I would like also to appreciate Fangming Xu who had helped me to implement the graphical interface of the active node system within the scope of his thesis.

I am deeply grateful to my colleagues of the National Laboratory of Switching Technology and Telecommunication Networks, Beijing University of Posts and Telecommunications, particularly to Prof. Shiduan Cheng, Meilian Lu, Yuehui Jin and Prof. Wendong Wang. Due to their help, I was able to fully concentrate on my study in Germany.

This work was supported by the German Academic Exchange Service (DAAD). I am very grateful to DAAD for giving me the opportunity and the financial support to finish this dissertation.

Finally, I would like to thank my parents and sister for their love and support in every way throughout my academic career and all those friends who made my five years in Germany enriching, memorable and fulfilling. Special thanks go to Ralf Behrens, who has been encouraging and helping me all the time during this work.

Abstract

Active networks enable the injection and execution of user-specified programs within the network infrastructure in order to support rapid deployment of new network services and protocols and to make the current Internet more flexible. Existing research has addressed the design and evaluation of the execution environments and the underlying operating systems of active nodes. Resource management has drawn attention for the security issues only, such as to prevent user programs from exhausting the system resources and to isolate processes created by the user programs in the active nodes. The provisioning of resource guarantees for the execution of user-specified programs in the active nodes, however, has been neglected.

Due to the introduction of user-specific computation in the active nodes, resources such as CPU and memory used by applications cannot be neglected like normally done in the current IP-based networks anymore. Moreover, due to the diversity of user programs, the usage of different types of resources in the active node systems may be non-balanced. Thus, situations that a user-specific computation cannot be executed due to the lack of only one or several types of resources may appear. Therefore, the resource management methods employed in the current Internet cannot be directly used in active networks. A more elaborated resource management method is needed in active nodes. The usage of each type of resources, the possible tradeoff among the resources, as well as the total system resource utilization must be considered.

In this work, a new resource management mechanism for active nodes has been developed. In order to address both the type and the amount of resources in active nodes, the concept of resource vector is introduced to facilitate the description of the resource usage in the active node system. Based on this, the adaptable resource vector space (ARVS) is suggested to describe the resource adaptability of applications. The ARVS formulates a generic model for network adaptation, and expresses also the individual resource requirements of applications.

An adaptive admission control mechanism has been designed to support the resource adaptation among different types of resources and among different applications in the active nodes. During the resource adaptations, it takes both the performance of the applications and the resource utilization of the node system into account. This mechanism supports both the generic network admission control and the flexible application-specific adaptation in the network nodes.

An active node architecture supporting the suggested adaptive admission control mechanism has been implemented. The resource management subsystem in the architecture realizes the admission control algorithm and performs resource adaptations. It oversees and manages the entire resource usage within an active node. Several experimental applications have been developed on top of this architecture, and used to evaluate the performance of the implemented node system.

Contents

List of Figures

Chapter 1

Introduction

1.1 Motivation

The IP-based Internet has traditionally been a data network and has in recent decades evolved into a well connected and widely distributed network like the telephone network. The IP-based network technology has been widely accepted due to its unparalleled ability to provide ubiquitous access and low prices. Nowadays, with the convergence of voice and data communications, it has emerged as a common medium supporting different forms of communications, such as distributed multimedia applications, streaming audio, video data etc. This has led to an increasing demand on the IP network to provide new network functions and services, and to be able to integrate new technologies and standards into the network infrastructure rapidly and easily. Traditionally, the development and deployment of new services or standards involves the standardization of requirements and capabilities, followed by thorough testing of the implementation prior to deployment. Therefore, the introduction of new services and standards is tedious and time consuming. It is not highly improbable that today's new services and applications may become obsolete by the time a standard for them is evolved. The essential reason is that in the traditional network the interface presented to users deals only with data rather than control or customisability. Users have no way to extend or customize the processing on a node within the network. Although in many modern switches and routers the service provider of the node can upgrade the software on the node, such modifications generally occur over a relatively long timescale only.

The active network (AN) approach has been originally proposed as an architectural solution for the fast and flexible deployment of new network services, in order to make the network able to support dynamic control and modification of its behavior. This proposition has ever been prompted by the thought to replace the numerous ad hoc approaches for performing user-specific computations within the network required by some applications, such as web proxies, multicast routers and mobile proxies. The basic idea of AN is to enable end users, operators and service providers to inject application-specific services in the form of programs into the network. Therefore, through the computations at the nodes within the network, users are able to utilize their specific services to obtain the required network function, and the new network services and standards can be incorporated on-demand.

Generally, AN represents a new approach to network architecture. The network is "active" in two ways: routers and switches within the network can perform computations on user packets flowing through them and may modify the packet contents; and users can "program" the network by supplying their own programs to perform these computations [TGS+96]. For example, users could request a router to execute an application-specific compression algorithm during the processing of their packets. In other words, AN supports dynamic modification of the network's behavior as seen by the user. The principal advantages of basing the network architecture on the exchange of active programs, rather than passive packets are (i) the exchange of code provides a basis for adaptive protocols, enabling richer interactions than the exchange of fixed data formats, (ii) application-specific functions can be placed at strategic points in the network, (iii) user-driven customization of the infrastructure allows new services to be deployed at a faster pace than can be sustained by standardization processes. Besides these, some hard and/or complex network problems such as the network management, traffic control and security are also expected to be solved better by using the active networking technology.

To realize AN, mechanisms that support users to inject customized programs into shared network systems are needed. These mechanisms involve the synthesis of work in programming languages, operating systems and networking. They can be organized in terms of enabling technologies, such as languages and compilers for "active" software; platform development, which leads to AN nodes suitable for deployment; programming models, with emphasis on interoperability across platforms; active controls and algorithms, which facilitate network configuration etc.; middleware services and applications, which can demonstrate the capabilities of AN. In short, realizing and using AN safely and efficiently faces many challenges. Efficiency, mobility and safety are among the central issues that must be considered and have to be compromised to some degree. Hence, significant work in the field of AN are focused on how to provide an efficient computation model and node architecture, how to solve the performance and security problems faced by the AN nodes and user programs. However, resource management in AN nodes has not been addressed sufficiently.

Nevertheless, resource management in AN nodes is necessary and becomes more difficult in comparison with in passive network nodes. On the one hand, at least in the near feature, there will not be abundant resources in the AN node systems for the arbitrary user-specific programs. But the injection and execution of user-specific programs in AN nodes demand for more types and larger amount of resources than the simple forwarding of packets in the traditional network nodes. Therefore, the available resources in an AN node system must be managed and scheduled in such a way that they can serve as many users and applications as possible.

On the other hand, the introduction of user-specific computation in the network results also in large changes on the resource usage in the AN node system. Both CPU cycles and memory are necessary for the user-specific programs, and they are not related to the network bandwidth as in the traditional network anymore. Hence, both of them cannot be neglected, but have to be considered explicitly in the AN nodes together with the network bandwidth. Moreover, with the increasing maturity of the AN technology and the AN infrastructure, various applications will be introduced and deployed in the AN. Since the features of user-specific programs are not known in advance, the resources in each AN node are exposed to users in ways not foreseen by the network providers. Furthermore, due to the large diversity of the applications and services the consumption of different kinds of resources in the AN nodes is not equally balanced: some applications prefer more communication resource, while others need more computation or memory resource or both of them. Generally, the ratio of the different kinds of resources can be configured appropriately according to the statistics of applications running in an AN node in the long run. However, in a short time, the status of the different kinds of resources in an AN node may be greatly diverse at different time points. E.g., an active application may require to execute a data compression algorithm in an active node, hence, needs large number of CPU cycles. When in a short time many such applications arrive at an AN node, the AN node will face the problem of no CPU resource available, although there may be still much available bandwidth resource in the node system. Therefore, it is possible that an application or services cannot enter an AN node due to the lack of only one type of resource. This poses a challenge on how to manage the multiple types of resources in a node in order to improve the total system resource utilization and satisfy the need of applications as much as possible.

Therefore, a proper resource management mechanism is needed in AN which takes the unique resource usage characteristics in AN into account. Resource management has been widely discussed in the traditional IP-based network in order to provide quality of service (QoS) guarantees for applications in the best-effort network environment. Some resource management models and architectures have been developed. However, most of the work deals only with the network bandwidth resource depending on the features of the traditional network and the needs of the applications. Several active node architectures related to resource management have also been implemented recently. However, they address resource management from the security point of view only, such as to isolate processes created by the user programs in the network nodes, or to limit the resources that an application or a packet can use in order to prevent them from exhausting the system resources. The provisioning of resource guarantees for the execution of user-specific programs in the AN nodes has been neglected. Resource management concerning the relationship among different kinds of resources and the system resource balance has rarely been studied. This prevents the system resources in AN nodes from being well utilized as well as serving the active applications, and may hinder the development of the AN and its applications.

More generally, this motivates us to study the resource usage and management in AN nodes from the perspective of the resource usage characteristics of active applications and the relationship among different kinds of resources. Because the resource status in the AN nodes changes rapidly over time due to the arrivals and departures of applications, as well as the changes of the application resource consumptions and possible resource failures and restorations, proper adaptation of the amount of resources obtained by each application can improve the performance of the whole system, including network nodes, end-systems and applications. Therefore, we adopt the resource adaptation mechanism to manage the resources in the node system. We perform the system-side resource adaptation, which means the AN nodes decide when and what kind of adaptations should be done. This can eliminate the complex and time-taking control in applications, and can also support the dynamic application requirements.

Therefore, our goal is to realize an adaptive resource management mechanism in AN nodes, which can provide resource guarantee for the execution of active applications on the one hand, and can, on the other hand, also keep the node system resources as balanced as possible, so that each type of the system resources can be better used and as many applications as possible can be served.

1.2   General Characteristics of Active Applications

Active network applications have certain characteristics which make reasoning about resource usage in AN difficult. As part of active applications, programs may be carried to the AN nodes. As a result, the content of packets can be changed in the AN nodes, new processes may be spawned, and new packets may be generated. More types of resources must be considered and large amount of resources can be consumed in the AN nodes in an unforeseen way.

In addition, applications are also flexible. The purpose for resource consumption of an application, in the final analysis, is to realize an appointed goal and achieve a certain performance. It is, however, often not important how this goal is achieved and which resources are consumed. For example, the data of a videoconference can be transmitted directly and can also be first compressed more highly and then transmitted further. Both methods can meet the need of the user but they consume obviously different bandwidth and computation resource. Another example is that different methods can be used to present certain data structures, such as sparse matrices and hashtables, where the well-known tradeoff exists between storage space and element access time. In other words, different amount of resources can usually be accepted by the applications.

These unique characteristics of active applications make it difficult both to describe the possible resource requirements and consumptions of active applications, and to manage the resource usage in AN nodes. Hence, not only a new method for describing the flexible resource requirement of active applications needs to be presented, but also a corresponding resource management method which takes into account both the characteristics of applications and the system resource balance must be proposed.

1.3 Problems to be addressed

In general, due to the complexity and flexibility of active applications, the following problems must be solved when introducing an adaptive resource management mechanism into the AN nodes from the perspective of active applications and the relationship of different types of resources.

First, an application may need more than one type of resources to finish its task in the AN nodes. Therefore, an approach for the description of resource requirements by the demanding applications is needed. A list of resource types and the corresponding amount can neither present the complexity of the applications in general, nor describe the flexibility of the applications. Hence, more elaborated schemes are necessary.

Second, the system-side adaptation can adapt the amount of resources allocated to an application very quickly depending on the changes of the resource status in the node systems. However, in order to realize the application-specific adaptations, these systems must be provided with some detailed information about the applications, especially their adaptation capabilities, namely, the resource demand scope within which adaptations can be performed and the execution result of the applications which can still be accepted by their users. Therefore, a way to describe and convey this information must be established.

Third, when and how to make appropriate adaptation decisions and to perform the adaptations in AN nodes are the key problems. This requires that the node system has to understand the resource requirement and adaptation preferences of applications. Furthermore, the system resource status must be well known and the resources in the system must be precisely controlled. In addition, the applications running in the node system must be monitored, so that they will not violate the resource agreement signed dynamically between the applications and the system.

Fourth, the adaptation should be done in a way not affecting the performance of applications and the node systems. The adaptation algorithm itself should not introduce much overhead to the applications and the node system.

And finally, in order to guarantee the execution of active applications in the AN nodes, the applications can only enter the node system if sufficient resources are available. This requires that an admission control mechanism have to be implemented in the AN node systems.

1.4 Our Focus and Approach

The existing resource management approaches in both the traditional passive network and in AN have not stressed the characteristics of resource usage in AN systematically.

They do not have a proper mechanism to solve the corresponding problems that may emerge, such as the description of the resource usage of active applications in AN nodes, and the resource balance among different resource types. This prevents the non-abundant system resources from being well used and satisfying the needs of users, and may potentially limit the development of AN.

This dissertation studies the resource management in AN nodes from the perspective of the relationship between different types of resources. It argues that multiple types of resources in AN nodes must be treated explicitly and the consumption of different kinds of resources in AN nodes is often not balanced. The resource-vector is introduced as the method for the description of the resource usage in AN nodes, and the adaptable resource vector space is suggested as the general approach to depict the adaptation capability of active applications. An adaptive resource management mechanism for active node systems is presented, taking the adaptability of active applications and the complementarity of different kinds of resources into account. The suggested resource management system performs admission control on new arriving applications, during which different steps of resource adaptations may occur depending on the resource status in the node systems and the adaptable resource vector space of each active application. Thus, the system resources can be adjusted among different resource types and re-distributed among different applications. In addition, an active node architecture with an explicit resource management subsystem which realizes the presented resource management mechanism is also implemented. Besides providing the general active network functions, such as supporting the injection and execution of application-specific code, this node architecture can also monitor and control the execution of each application according to the result of the dynamic adaptive resource admission control. Evaluations about the adaptive resource management mechanism are also given based on the implemented AN node architecture.

1.5   Outline

The remainder of the dissertation is organized as follows:

Chapter 2 provides background information for understanding this dissertation and its achievements. The main principles of the traditional IP-based network are summarized, followed by the discussion of the unique features of AN and the key issues involved in AN, including the level of programmability, the general architecture of AN nodes, and the basic approaches for developing applications in AN. In addition, an overview about other technologies that make networks programmable and the general resource management methods currently used in the traditional IP-based network are also presented. Moreover, the related research work about resource management in programmable networks is surveyed. As a summary of this chapter, the differences between the existing work and our research focus are addressed.

Based on a detail analysis of the resource usage in AN nodes and the characteristics of resource consumption of active applications, chapter 3 introduces the concept of resource vector, which is used to describe the resource usage in AN. Two categories of operations on resource vectors are introduced. By using the resource vector concept and the operations, not only the multiple types of resources needed by an active application can be expressed, but also the possible relationship among these resources, which has been neglected in current work, can be emphasized. In addition, the concept of resource vector suggests also the method for allocating the system resources while keeping the usage of different types of resources in the system in balance, and provides a basis for the description of the adaptability of applications while taking their flexibility related to resource usage into account. These issues have not been considered by other work so far.

Chapter 4 presents the adaptable resource vector space (ARVS) to describe the resource adaptation capability of applications after an analysis about the adaptability of resource consumption. The components used to define ARVS are introduced and discussed. ARVS stipulates the scope for the node system to adjust the resources that can be used by applications without affecting the acceptance of applications. Based on the components of ARVS, the adaptability information of an application can be transferred to the network in an integrated manner. This solves the major problem for realizing the application-specific adaptations in the network nodes, and at the same time, the flexibility of applications related to the resource usage are also considered. By using ARVS, adaptations can be performed in the node systems according to the requirement of each application, and the complexity and the time-taking problem caused by integrating the adaptation mechanism in applications can be reduced. In addition, work related to the description of application adaptability is also discussed in this chapter.

A detailed description of the adaptive admission control mechanism introduced in the AN nodes is presented in chapter 5. Unlike other works till now, this mechanism is based on the resource vector and ARVS, and introduces a resource adaptation algorithm during the admission control phase. The system resources can not only be adjusted among different resource types, but also be re-distributed among different applications. This mechanism facilitates the resource balance in the network node system, and can improve the total system resource utilization through serving as many applications as possible. In this chapter, an overview of the related optimization techniques is given first, and then the resource adaptation algorithm is elaborated. Performance evaluation related to the algorithm and related work are also presented.

Chapter 6 discusses the implementation of the active node system which realizes the proposed adaptive admission control mechanism. The most distinguished feature of this node system is that it implements an explicit resource management subsystem which carries out the adaptive resource admission control mechanism, monitors and controls the different kinds of resources used by the system and each application in the system. In this chapter, the architecture of the active node system is described, the fundamental components realizing the basic node functions and the resource management subsystem are depicted, and the mechanisms for monitoring and controlling each resource type are elaborated. Performance evaluation related to the implementation and related work about the resource accounting are also discussed.

Chapter 7 evaluates the implemented AN node architecture with adaptive admission control mechanism in terms of the introduced overhead, the improvement to the system resource utilization and affects on applications in the test AN constructed on the practical Internet. The evaluation demonstrates that the proposed approach is feasible and comes with acceptable overhead. Applications developed on top of the AN node system and used to validate the admission control algorithm are also introduced.

Finally, chapter 8 summarizes the main achievements of the dissertation, and gives an outlook about potential future work.

Chapter 2

Background and Related Work

This chapter gives an introduction to active networks (ANs) and resource management in general. Furthermore, related work in the field of this dissertation is surveyed. Overall, the purpose of this chapter is to provide sufficient background information for the understanding of this dissertation and to see the achievements of this dissertation.

2.1 Traditional IP-based Networks

In this work, IP-based networks are regarded as the collection of networks communicating through the use of the Internet Protocol (IP) [Pos81]. Such networks consist typically of routers responsible for forwarding packets and end-systems that send packets to each other. All these nodes are interconnected by various types of network links.

Traditionally, the development of IP-based networks was guided by the principles of the end-to-end argument [SRC84], which suggests that a function or service should be carried out within a network layer only if it is needed by all clients of that layer, and it can be completely implemented in that layer. Following this argument, supplying complex functionality within the network is sub-optimal, since it is probable that such functionality will either duplicate the work that the application needs to carry out at a higher level, or will fail to meet the needs of most users. Hence, IP supports an unreliable unsequenced form of data delivery only. The routers process packets in a connectionless store-route-forward model, forwarding packets from a source host to a destination according to certain fields in the header of the packets. Since the resources consumed during storing, routing and forwarding are related to the number and length of the packets, typically only network bandwidth resources needed by the packets are considered in the traditional IP networks.

The traditional IP-based networks are best-effort networks. I.e., the network infrastructure itself cannot provide any quality of service (QoS) guarantee for the applications. QoS has to be provided through special techniques. The QoS provisioning issues have been addressed in several ways, two of the primary solutions being developed for the traditional IP-based network are:

9

- Integrated services, which allow each network flow or class of flows to be mapped to a particular QoS class. This QoS class is used to schedule the transmission of packets in these flows. The resource reservation protocol (RSVP) has been proposed as a protocol for the exchange of flow specifications among routers.

- Differentiated services, which attempt to reduce the amount of state required within the core routers by requiring end nodes and intermediate routers to mark each packet with the required forwarding behavior. Service Level Agreements (SLAs) may be set up between different network domains specifying the traffic profile for different classes of packets.

In conclusion, the traditional IP networks are based on a layered architecture. The interoperability is acquired through standardizing the syntax and the semantics of the data frames of each layer. In other words, the essential feature of the traditional IP networks is that the interface presented to users deals only with data rather than control or customisability: the users have no way to extend or customize the processing on a node within the networks. Furthermore, although in many modern switches and routers the provider of the node can upgrade the software on the node, such modifications generally occur over a relatively long timescale.

2.2   Active Networks

The concept of ANs emerges from the discussions within the DARPA research community in 1994 and 1995, and one of the first descriptions is given in [TW96]. The basic idea stems from the identification of the problems of the traditional networks: the difficulty of integrating new technologies and standards into the shared network infrastructure, poor performance due to redundant operations at several protocol layers, and the difficulty of accommodating new services in the existing architecture model. The emergence of some network applications, such as firewalls, web proxies, multicast routers and mobile proxies, has prompted the advent of ANs. These applications require some form of computation within networks. But due to the absence of architectural support, these applications have to use a variety of ad hoc approaches for performing user-driven computations at nodes within the network. Hence, the idea comes up to replace the numerous ad hoc approaches with a generic method allowing users to introduce programs into networks, namely active networking. The goal of ANs is to speed up the evolution of network technologies through making the networks programmable and allowing users to introduce new protocols and services.

2.2.1   Overview

An AN consists of a group of network nodes, such as routers, where an environment is provided for the deployment and execution of user-specific applications. These nodes are called active nodes or AN nodes. In contrast to the traditional IP-based networks, which provide in fact a transparent processing for the data passing through them, ANs have two general meanings. One is that ANs can be viewed as store-execute-forward systems: after receiving the data from users, an AN node may performs some specified computation on it, and then forwards the processed data or new generated data continuously. The feature here is that the content of the user data can potentially be changed before being forwarded. The other meaning of "active" is that users are allowed to inject their own programs into the networks, which means the computation at the active nodes can be performed according to the dynamic requirement of the users. Namely the networks can be programmed by the users and the functionality of the network can be dynamically extended.

Realizing and deploying ANs face several challenges. They can be summarized in the programming model, safety and security, performance as well as management and control of the ANs.

The programming model involves the computational model, node architecture and end-to-end architecture. Currently the approaches for solving this problem can be summarized into four categories according to the level that the programmability is provided by each programming model.

1. **Capsules.** This approach provides the programmability at the most fundamental level. Each packet may contain both executable code and data. Such a packet is often referred to as a capsule. When a capsule arrives at an AN node, the code carried in the capsule is loaded into the execution environment provided by the node system and begins to run, typically using the parameters carried in the data field of the capsule. Possible actions performed by the capsule includes:

   - Performing a routing decision and transmitting itself forward.
   - Creating new capsules and sending them to the original or a new destination or back to the source of the capsule.
   - Interacting with objects or other states in the node system, in order to obtain information about the node or the local network, or to process itself depending on the information.
   - Performing some control and management operation according to the network status.

   A capsule may also carry some identification to specify the code that should be used to process itself instead of carrying code along with it.

   Typical systems using this programming model include ANTS and PAN from MIT, ALIEN from the University of Pennsylvania, Smart Packets from Kansas University and PLAN from the University of Pennsylvania and Bellcore. ANTS [WGT98] and PAN [NGK99] support the on-demand code loading and caching mechanism. Each capsule is tagged with a code identifier, identifying the code needed to process this capsule. If the code cannot be found in an active node, it can be loaded from the previous node dynamically using capsules. PLAN (Packet Language for Active Networks) [HKM+98] [HMA+99] is built around the two concepts of invoking the named services on a local node and a function on a remote node, where a remote invocation is semantically equivalent to sending a packet. A PLAN capsule contains a chunk, which represents a delayed remote PLAN evaluation. An ALIEN capsule consists of an OCaml [Ler97] bytecode module and a payload; and a SmartPacket [SJS99] contains the serialized class definition and states of a Java object.

2. **Extensions.** This approach provides the programmability at an intermediate level. To some extent, the active extensions can be considered as programs used to process the packets from users. In this approach, the extensions are loaded on to a router using an out-of-band mechanism. E.g., the Composable Active Network Elements (CANES) [BCZ98] architecture supports a limited form of programmable extensions. Within CANES the behavior of a flow of packets can be customized by selecting triggers that invoke predefined programs at certain points during packet processing.

3. **Partially Programmable Networks.** Active networking can also be realized by programming part of the network nodes. The active router control is a hybrid active/passive architecture that assumes a passive IP forwarding network as the main "transport place", and associates a "controller" with groups of IP routers to deal with high level issues such as maintenance of routing tables. The controller itself may be an active node, thereby permitting flexible customization of the network by users without degrading the performance of standard IP flows that have no requirement for active behavior.

   An example using such a programming model is the Active Networks Overlay Network (ANON) [Tschu99], which supports the creation of programmable clusters of active nodes connected together by a large overlay network.

4. **Active Services.** This approach provides the programmability at the highest level. The corresponding programming model permits the programmability within the networks, but only at the application layer. User-supplied code may be executed on nodes within the networks, but cannot replace the processing at the network or the transport layer. This approach is also referred to as the Application Layer Active Networking.

In these programming models, the higher the level where the programmability is provided, the less flexibility can be provided by the programming model.

Due to the provisioning of the programmability, ANs face also the safety and security challenge. ANs become more open to users. The arbitrary programs from end-users may change the states of the router, access more types and amount of system resources and cause the denial of service in the node system. This requires that the mobile programs should be safe. They should not be harmful to the AN nodes. Simultaneously, the AN nodes should also own mechanisms to protect themselves from attacks of mobile code with or without deliberateness. In addition, the AN nodes should also provide a secure environment for the execution of user programs, keeping privacy of one user program from other users. Currently, the main mechanisms used for providing a safe active node architecture include the language-based and the resource-based control.

The language-based control requires the end systems to use a safe language. Some systems approach this issue by using an existing safe language. For example ANTS [WGT98], PAN [NGK99] and SmartPackets [KMH+98] use the Java language in slightly different ways. ANTS and PAN process capsules using the dynamically loaded Java classes; SmartPackets capsules contain the full definition for a Java class as well as the serialized data for a single instance. Caml [Ler97] is a dialect of ML [MTH+97] that has been used by RCANE [Men99] and several components of the SwitchWare [AAH+98] architecture such as ALIEN [Ale98] and the Secure Active Network Environment (SANE) [AAK98]. Other systems have developed their own languages specifically designed for mobile computing. Among them, PLAN [IIMA+99] is a packet language for ANs developed by the SwitchWare project. It is a restricted functional language similar to ML, but with strong dynamic typing. The language restrictions allow PLAN programs to be safely executed without verification or checking. Sprocket is also a safe high-level language specially designed for ANs [SJS99]. It is mapped into a restricted CISC "assembly language", called Spanner, and a virtual machine environment in which the Spanner programs can be executed.

RCANE [Men99] and Janos [Janos] are examples of systems that protect the active nodes through controlling the resources used by untrusted codes. We will discuss these projects in section 2.4.2 and 6.1.1 respectively.

Performance is also a challenge for realizing ANs. Normally, an AN node has to create a separate execution environment for processing an active application. Spawning new environments and processing data using different programs, or creating new services mean that much more operations must be performed in AN nodes than in the traditional network nodes where usually only routing and forwarding operations are done. It is a common agreement that the performance of the ANs should be able to compete with that of the traditional IP networks. Works like SNAP [Moo02] and PAN, which are the follow-on projects of SwitchWare and ANTS respectively, stress the performance and efficiency of the AN node architectures.

The management and control of ANs as well as the resource management in ANs must also be considered in order to be able to deploy ANs appropriately and provide satisfactory services for users.

The benefits that can be gained from the programmability of ANs include:

- Rapid deployment of new network protocols. In the traditional IP networks, deploying new network-level protocols is a slow and expensive process. A new protocol might take over a decade to travel from prototype stage to large-scale deployment. Examples include RSVP, multicast services, security and mobility extensions and IPv6. By providing a programmable environment where users can define the custom packet processing routines, new network-level protocols can be deployed substantially more quickly.

- Fast incorporation of new services. By using the programmable environment, incorporating new services means simply executing the code dynamically injected by users. Therefore, it becomes much easier to incorporate new services in ANs than in traditional networks, and ANs are more flexible.

- Solving hard and/or complex network problems. Some hard and/or complex network problems are also expected to be solved better with the help of active networking technology. These problems include QoS routing, differentiated service provisioning and signaling, reliable multicast and multicast routing, network management and traffic control etc.

- User-specific data processing in networks. In ANs, users can inform the network nodes about how to process their data by making full use of the information in the networks. Examples include web caching, online auctions, mixing of sensor data, high-performance gateways and so on.

In the following, we present in more detail the general architecture of an AN node, the organization of the resources in it as well as the methods for developing applications in ANs.

2.2.2   General Node Architecture

A general architecture for active nodes [Cal99] is suggested by the AN research community in order to provide an environment for loading and executing the programs injected by users in the network nodes, and at the same time to retain enough common interfaces so that the programs from users can run on as many active nodes as possible. The general active node architecture deals with how the programs from users can be executed, and what kinds of and how the local system services and resources can be accessed by the user programs. In addition, how the local system resources in this architecture are organized is also suggested in [AN01].

As illustrated in Figure 2.1, the most essential functional components of an active node are the Node Operating System (NodeOS) and the Execution Environments (EE). The EE is in fact a virtual machine in which a user process is executed. It provides an interface for applications by exporting APIs. The resources and services provided by an active node can only be accessed by the programs from end users through the EE. The EE is also responsible for all other aspects of the user-network interface, such as the syntax and the semantics of the packets submitted by users, the nature of the programming model and the abstractions supported, as well as the addressing and naming facilities etc. Through EE, the independent implementations of applications can be executed in the active nodes.

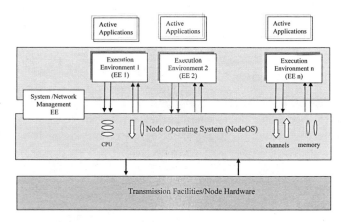

Figure 2.1: General architecture of active nodes

There may be multiple types of EEs in an active node. Applications can specify in their packets to the networks, which type of EE they use. Since the EE is essentially the programming environment of ANs, each type of EE is usually concentrated on a particular language or programming model. It accepts valid programs and packets, executes them and/or modifies the system states. However, the number of different types of EEs supported in an active node could be small due to technical reasons, e.g., the difficulty of development and deployment of a new EE, the communication among different EEs etc. Normally, the functions for dynamic downloading and installing of a new EE type are only accessible to the node administrators via the management EE, because they may put the network resources in a risk if the EE misbehaves.

The NodeOS provides support for the existing of multiple types of EEs in an active node simultaneously and safely. It is responsible for the isolation among different types of EEs and among different user processes running in one EE type. The states in one EE type should not be affected and changed by other types of EEs, and no user process can access resources and variables available to other processes. In addition, the NodeOS provides some basic functions that EEs can use according to the requirements of applications. The EEs interact with the NodeOS through the Node Interface [AN01]. Moreover, just like the OS of a general-purpose computing system, the NodeOS is also responsible for managing the resources in the active node and providing abstractions between user processes and the underlying system resources. Through the resource allocation and control, it separates the work of the EEs from each other. The NodeOS multiplexes and mediates the communication, memory and computational resources in the node system among the EEs created for applications. It also ensures that no single process captures too many resources to hinder the performance of other user processes or normal functions of the node itself.

In addition, the active applications (AAs) in the node architecture are the end user processes that run in the node system and require end-to-end communication or service. They take advantage of the functions and services provided by the node system.

2.2.3   Resources in Active Network Nodes

The programmability of the active nodes increases greatly the ways in which the programs of end users may consume the resources in the node system. Generally, the obvious resources at an active node are CPU cycles, memory and transmission bandwidth at the output physical links. Other resources include also:

- Cache memory, e.g., caching the program code used for processing a definite type of capsules.

- Persistent store, e.g., storing the coarse-grained sharable persistent states.

- Specific hardware, e.g., for encryption, DSP, and special links.

- Some specialized data, such as routing table entries.

The AN NodeOS working group has abstracted four primary concepts to describe the usage of resources in an active node [AN01]. They are thread pools, memory pools, channels and domains. The first three encapsulate the major three types of resources in the node system, i.e., the computation, storage and communication resource. And the fourth is used to aggregate the control and scheduling of the other three in a form that resembles the requirements of applications more closely.

- **Domains** are the primary abstraction for accounting, admission control, and scheduling in the system. A domain typically contains a set of channels on which messages are received and sent, a memory pool, and a thread pool. Active packets arrive on an input channel are processed by the EE using threads and memory allocated to the domain, and are then transmitted on an output channel. One can think of a domain as encapsulating resources used across both the NodeOS and an EE on behalf of a packet flow. A given domain is created in the context of an existing domain, making it natural to organize domains in a hierarchy, where the root domain corresponding to the NodeOS itself. The second level of the hierarchy corresponds to EEs and domains at lower levels are EE-specific.

- **Thread pools** contain threads that execute application programs and shepherd messages across the domain channels.

- **Memory pools** contain the buffers used to hold the objects of the application-specific programs and the EE-specific states, as well as queue messages on domain channels.

- **Channels** are used to send, receive and forward packets. Normally, each channel has some associated protocol processing such as IP, TCP, UDP, or physical transmission links such as ATM, Ethernet etc. When a packet arrives upon a physical link, the NodeOS classifies it according to the protocol type carried in the header and place it in the appropriate logical input channel. The channel delivers the packet either to an EE for interpretation and processing or to an output channel for transmission. Therefore, a channel consumes not only network bandwidth, but also CPU cycles and memory buffers.

An EE processes packets using the threads and memory allocated to the corresponding domain. New packets may also be generated during the processing and put on the output channels.

2.2.4 Active Network Encapsulation Protocol (ANEP)

An active node is capable of dynamically loading and executing user programs written in a variety of languages. However, it selects the environment for executing the programs according to the user requirements. The active network encapsulation protocol (ANEP) [ABG+97] has defined the general format of the active packets, which carry some instruction information in the header and the user programs in the payload. Figure 2.2 illustrates this format.

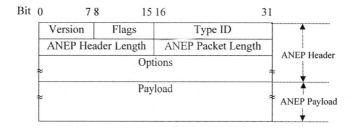

Figure 2.2: ANEP packet format

The Version field identifies the version of the ANEP. The Flags field indicates what the node should do if it cannot recognize the Type ID field. It is recommended now that the value is set to 0 by the packet originator, which means the node should try to forward the packet using the default routing mechanism, if the necessary information is available in the Options part of the header. If the value is 1, the node would discard the packet.

The Type ID field is used for selecting the type of EE in which the program should be executed. Since there may be multiple types of EEs in an active node, an application should notify the active nodes to route the packets belonging to it to a particular type of EE. Now in ANs, some values of Type ID have already been assigned to different types of EEs. If the specified EE type is present at a node, packets containing a valid ANEP header with the corresponding Type ID will be routed to channels connected to the indicated EE. If the specified EE type is not present, the packets will be routed to the default EE type or be discarded according to the indication in the Flags field. The proper authority for assigning Type ID values to interested parties is the Active Networks Assigned Number Authority (ANANA).

In addition, there is also an Options field in the ANEP header. Various options can be specified by the applications, such as authentication, confidentiality or integrity. Applications can implement or specify methods in the active nodes telling them what is in the Options field and if and how they should be processed. More information about the Options field can be found in appendix A.

How to transfer the ANEP packet through the network is node-dependent. Currently, most of the EE implementations pack the ANEP packets in the UDP datagram, such as ANTS and PLAN etc.

2.2.5   Applications in Active Networks

Active applications accomplish some desired communication functions using a combination of packet forwarding and computation in the network nodes. An active application may operate in the data plane, processing data packets in real time; or operate in the control or management plane, setting or querying control states. Active applications are basically written in some constrained languages that provide portability and security. Their execution requires both the network and the node system resources such as CPU time and memory.

In general, depending on the level of the programmability provided by the AN nodes, applications may adopt two approaches to program the networks and realized their demanded network functions.

Integrated approach: also widely known as the in-band or the encapsulation approach. In this approach, the application-specific programs are divided into one or several fragments and encapsulated with other user data into one or several active packets, usually called capsule(s), and sent to the network. In other words, an application packet may contain a program or a program fragment; each program or program fragment may or may not have some embedded data. The program is composed of instructions that perform the basic computations, and can also invoke some "build-in" primitives or functions, which may provide access to the resources external to the transient environment where the computations are executed. Using this approach, the applications do not request to be processed by a protocol, but rather carry the protocol themselves in the form of program as they travel through the network. In this case, the applications are relative complex but very flexible. This method requires that the underlying AN nodes must support the programmability at the capsule level.

Discrete approach: this is also called the out-of-band or the programmable node approach. In this approach, application programs are injected into the active nodes separately from the actual data packets traversing through the network through an auxiliary mechanism. I.e., active nodes contain already a number of predefined protocols. Application packets need to specify one of the already installed protocols to process the carried data. Compared with the integrated approach, the discrete approach simplifies the applications. However, using the discrete approach the application-specific functions cannot be directly and immediately executed. In this case, the programmability of the AN nodes can be provided in the extension or higher levels.

By now the applications developed for ANs include dynamic network congestion control [BCZ96], downloadable diagnostic functionality of network management [RS99], online stock quotes [WLG98], detection and automatic mounting of barriers to web viruses and attacks [SDB+02], multicasting [SKB+01] etc. They take advantage of the programmability provided by ANs for network caching, data fusion, data filtering and data detection. Possible active applications include also those which process flows within the network, such as to add FEC (Forward Error Correction) at lossy points within a network, to transcode a high-bandwidth multicast stream into a lower bandwidth stream before sending it over a low- bandwidth link, customized routing etc.

2.2.6   Other Programmable Technologies

Besides active networking, the open programmable interface networks technology aims also to make the networks programmable. Furthermore, since new value-added services can also be provided by using the mobile agent technology, we believe that the mobile agent technology devotes also to make the networks programmable. Figure 2.3 summarizes the approaches for providing programmability in networks. Generally, the differences between ANs and the other two techniques lie mainly in where and how to realize the programmability, as well as the level of the programmability. In the following we take a brief look at these two techniques, concentrating on the differences between these techniques and ANs.

2.2.6.1   Open Programmable Interface Networks

The Open Signaling (OpenSig) community [OpenSig] advocates that the programmability can be achieved by means of defining a series of open network interfaces that represent physical network devices and network services as distributed objects. Therefore, this approach is referred to as the open programmable interface networks technology. Analogously to ANs, who try to open up the routing and the packet processing functions on

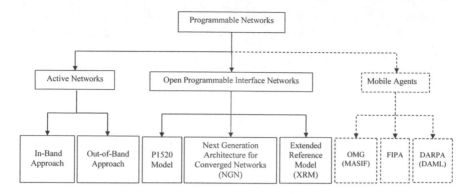

Figure 2.3: Approaches providing programmability

the data path of a network node, OpenSig has attempted to provide a programmable interface for the control plane of a network node, typically a router, an ATM or telephony switch, enabling access to internal states and control of network devices. Thus, independent software vendors and service providers can enter the telecommunications software market, thereby fostering the development of new and distinct architectures and services.

The IEEE P1520 standardization effort [BLM+98] is the leading industry initiative for standardizing programming interfaces for networks to enable rapid service creation and open signaling. It envisions a network as a giant computer, a fully programmable machine that delivers advanced voice, data and video services globally. It opens up application programming interfaces (APIs) for the network, just like an API provided by a general computer.

Figure 2.4 illustrates the P1520 reference model. The model is a layered architecture. It consists of 5 levels with 4 interfaces. Each level comprises a number of entities in the form of algorithms or objects representing the logical or physical resources depending on the scope and functionality of the level. The interfaces offer services to the level above it and abstract the components below it for customization or programming. In more detail, the V-interface (Value-added services interface) provides access to the value-added services level. This interface provides a rich set of APIs supporting highly personalized end user applications through value-added services. The U-interface (Upper interface) deals with generic network services. It creates a separation between the actual interface and vendor implementations, allowing multiple network-level schemes to coexist in a single network. The L-interface (Lower interface) provides software abstractions for both the value-added services and the network generic services level. It defines the API to directly access and manipulate local device network resource states. The CCM-interface (Connection Control and Management interface) is a collection of protocols that enable the exchange of state and control information at a very low level between a device and external agent, namely through the CCM-interface, the network physical elements can be accessed.

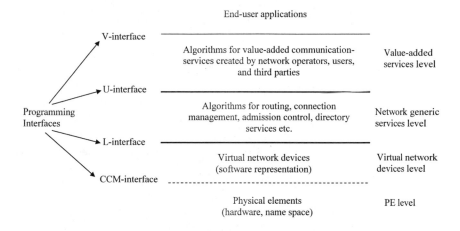

Figure 2.4: IEEE P1520 reference model

Currently, this technique has been pursued to

- Enhance internetworking functions as needed, for example, translation between media protocols, such as Internet and the existing intelligent network-based telephony or different signaling systems like SS7 and H.323.

- Manage the network more powerful than possible with the standard SNMP (Simple Network Management Protocol).

- Implement traffic control algorithms to support QoS.

The Next Generation Network (NGN) [MU01] architecture for converged networks uses the concept of open programmability to provide a solution that converges the services provided on the Internet and the Public Switched Telephone Network (PSTN). The Extended Reference Model (XRM) models the communication architecture of networking and multimedia computing platforms. It consists of three components, the Broadband Network, the Multimedia Network and the Services and Applications Network. Xbind [LBL95] [LLM96] is the typical work based conceptually on XRM.

Generally speaking, the programming model advocated by ANs is more radical than that advocated by OpenSig. ANs allow the customization of network services at a packet granularity. Techniques like expressing languages, compiling system and execution environment are involved in ANs to provide a dynamic runtime support for packets carrying user programs. OpenSig provides a programmable control plane with an open API only. However, the scope of activities in IEEE P1520 covers technologies from ATM switches to

IP routers, and to media gateways. It separates clearly network control from information transport. ANs primarily focus on IP networks, where the control and data transmission are combined.

2.2.6.2 Mobile Agents

Mobile agent technology is another research area related to network programmability. It emerged from early work done in the '80ies on distributed and mobile objects, process migration, and distributed artificial intelligence. Nowadays, there is a large amount of literature on this subject, applied to several areas such as manufacturing, e-commerce and network management. The main advantages of mobile agent have been identified such as software distribution on demand, asynchronous operation of tasks, reduction of communication cost, scalability due to dynamic placement of functions etc. The agent standardization efforts include the MASIF (Mobile Agent System Interoperability Facilities) from OMG (Object Management Group Agents Working Group) [MBB+98], FIPA (Foundation for Intelligent Physical Agents) [FIPA] and DAML (DARPA Agent Markup Language) [DAML].

Figure 2.5 illustrates the general agent model. An agent system consists of a number of abstract places, which are the execution environments where agents live. Agents are entities. A system agent is a specialized agent allowed to access the system resources of the host of the agent system. A mobile agent is an executing program that can migrate during execution from one host to another in a heterogeneous network. The states of the running program are saved, transported to the new host and restored, allowing the program to continue where it left off. On each host, the agent interacts with stationary service agents and other resources to accomplish its tasks.

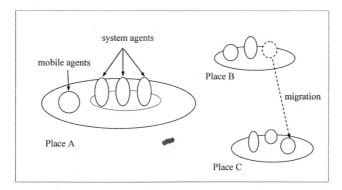

Figure 2.5: The agent model

The mobile agent technology can make the network programmable in the way that new network protocols or value-added services can be packed into the mobile agents.

Through the execution of the mobile agents in the network nodes, the functionality of the networks can be extended. To some extent, it can be considered that some pioneer AN platforms came from a mobile agent framework: the M0 platform [Tschu93]. M0 is based on the concept of messengers, which are mobile computational entities that are able to perform any network services. In [Tschu93], it is shown that any protocol based on the classical PDU paradigm (which is the case of the most network protocols in operation so far) can be implemented using the messenger paradigm. It is also pointed out that not all the protocols that can be implemented by the messenger paradigm can be implemented using simple PDUs: the typical case is the one that some protocols are able to evolve their own code on the fly.

The intersection between the mobile agent and the AN technology is the use of mobile code. To some extent, the capsules of an AN can be seen as subclasses of mobile agents specialized for network related operations. Conceptually there is no clear border line between mobile agents and capsules in ANs. But in practice, some differences still exist between the two approaches. The architectures for mobile agent platforms tend to concentrate on application-level or network value-added services, while AN platforms are optimized for transport rather than processing of information. In addition, mobile agents focus mainly on application-level or network management duties, and can typically accomplish much more complex tasks with richer functionality than what is generally allowed to capsules. These make also that the resource management issues are more critical in AN than mobile agents since AN is closer to the node system level.

On the other hand, mobile agents can also be used as the enabling platform for ANs as pointed in [SMY+99]. In this case, it is possible to benefit from full mobile agent functionality in an AN environment. [BCL99] has presented an agent-based AN architecture where two example applications demonstrate that mobile agents can enhance the robustness and performance of the AN solutions.

2.3   Resource Management

2.3.1   Overview

Resource management has been widely studied in the field of computer systems and networks, where the system resources, such as CPU, memory, various I/O systems, network bandwidth etc. are shared among multiple tasks and applications. It is based on the fact that the resources are not abundant. This remains true due to the ongoing increase of applications in both quantity and quality, although with the improvement of technologies the resources in the systems become more and more. In this section, we concentrate our discussion on resource management in the field of networks.

Generally, two issues have motivated the studies of resource management. One is safety and security, namely to protect the system by limiting and controlling the resources used by each application, and thereby also providing fairness and protection among applications in the system, such as the Resource Containers [BDM99] and the Resource Controlled Active Node Environment (RCANE) [Men99]. The other is to provide QoS support for applications. I.e., the limited resources must be managed and scheduled as well as possible so that they are available when needed, in order to meet the needs of as many applications as possible. Examples include Rialto [JRR97], Eclipse [BBG99], Nemesis [Nem], and [NS95].

In the context of providing QoS support for applications, the requirements to a resource management mechanism includes:

- Satisfying the desire of each application as well as possible. Each application has its own performance requirement in such a resource sharing system, which is typically represented by the QoS parameters, such as delay, delay jitter etc. By managing the resources in the system, the performance requirement of applications or tasks should be satisfied as well as possible.

- Keeping fairness among applications or tasks. A resource management mechanism should also take the fairness among the tasks and applications in the system into account. Normally, the tasks or applications are assigned some priorities by the system to indicate their importance. Tasks or applications having the same importance should be treated by the system equally.

- Improving the performance of the whole system, such as response time, CPU utilization or throughput. In other words, the system resources should be best utilized. Depending on the function of systems, the criteria for evaluating the system performance may be different.

Hence, the basic tasks for resource management involve:

- Calculating the resource requirement of each application: the resource management system has to know first how many resources each application needs. This can be obtained through translating the QoS requirements of applications or directly from the applications.

- Allocating resources to each application: the resource management system has to reserve the resources needed by the applications. Note that in this case the reservation can be physically or virtually (just marking).

- Deploying the resources: scheduling the resources among multiple applications according to some scheduling algorithm. In this phase, the resources are consumed by the applications and the applications acquire their expected performance correspondingly.

Among these issues, scheduling is the most basic one. It decides ultimately how many resources an application or a task may acquire. Through scheduling, the executable tasks or applications are rationally mapped onto the executing units, which occupy the system hardware or software resources. In the network system, since the transmission bandwidth is shared among flows or connections, the transmission scheduling algorithms have always to take the fairness among the flows or connections into account, such as weighted fair queuing [DKS89], virtual clock [Zhang90] and the modified form of EDF [BBD+97].

Besides for providing QoS support for applications, the system resources must also be managed for the purpose of safety and security. Resources used by each application must be limited in order to prevent it from affecting other applications or the system by overusing or exhausting the system resources. Some mistakes of application programmers may result in the overuse of system resources and affect or block other applications in the system. Some malicious applications may attempt to "flood" a network, thereby preventing legitimate network applications from executing properly. Particularly with the increase of the flexibility of networks, the networks become also more open. The applications can access more resources of the underlying network systems. Thus, the possibility that the systems are attacked deliberately or non-deliberately becomes also greater. Some work about controlling the resources consumed by each application has been done in the context of resource management for protecting the systems, especially in the field of mobile code, active and programmable networks. Such work includes Janos, Rcane etc. We will discuss them in detail in section 2.4.2 and 6.1.1.

In the following, we present two typical resource management techniques used in the current IP networks for providing QoS guarantees for applications, namely resource reservation and adaptation.

2.3.2   Resource Reservation

Resource reservation means that the network nodes set aside certain resources for a user flow that has quantitatively required a QoS. Typically, the QoS provisioning through resource reservation is implemented with the help of a reservation protocol such as RSVP [BZB+97] or YesSir[PS98], and traffic control mechanisms within the network nodes. The reservation protocol is responsible for the creation and maintenance of resource reservations on each link along the transport path. It is used by a host to request specific QoS from the network for particular data flows, or used by routers to deliver QoS requests to all nodes along the paths of the flows and to establish and maintain states to provide the requested service. The traffic control mechanisms are responsible for providing the promised QoS.

The integrated services (Intserv) [BCS94] model is a known representative providing per-flow QoS guarantees based on resource reservation. The traffic control mechanisms consists of packet classification, admission control, and packet scheduling or some other link-layer-dependent mechanisms that determine when individual packets must be forwarded. Admission control is required to ensure the node has sufficient resources available to meet the QoS demands of each new request. The packet classifier determines the

QoS class for each packet based on its flow affiliation, and sorts packets into different treatment groups. For each interface, the packet scheduler or other link-layer-dependent mechanisms are responsible for forwarding packets in different ways according to the promised QoS. The resource reservation protocol (RSVP) is the widely known protocol used in IP-based networks.

Figure 2.6 illustrates the functional modules of RSVP in routers. During the reservation setup, a RSVP QoS request is passed to two local decision modules, Admission Control and Policy Control. Admission Control determines whether the node has sufficient available resources to supply the requested QoS. Policy Control determines whether the application has the administrative permission to make the reservation. If both checks succeed, the parameters for obtaining the desired QoS are set in the packet classifier and the packet scheduler. If either check fails, the RSVP program returns an error notification to the application that has originated the request.

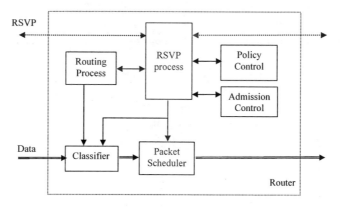

Figure 2.6: RSVP in Routers

Generally, RSVP makes reservations for unidirectional data flows. It is receiver-oriented, i.e., the receiver of a data flow initiates and maintains the resource reservation for that flow. The result of a successful RSVP request is that the resources in each node along the data path are reserved.

The disadvantage of the Intserv model is that the use of per-flow state maintaining and processing raises scalability problems for large networks. This has resulted in the appearance of the class-based differentiated service (Diffserv) [BBC+98] model, which offers services on an aggregate basis rather than per-flow and forces as much per-flow states as possible to the edges of the networks [BYF+98]. The service differentiation is achieved by mapping the codepoint contained in the Differentiated Services (DS) field in the IP packet header [NBB+98] to a particular forwarding treatment, or per-hop behavior (PHB), at each network node along its path. [WJK+02] has presented a framework of resource management in Diffserv (RMD) designed for the edge-to-edge resource reservation in a

Diffserv domain.

2.3.3   Resource Adaptation

Another technique for managing resources is adaptation. Basically, resource adaptation means to dynamically adjust the resources allocated to applications. Two observations in the networks have directly resulted in the research of resource adaptation: one is the available resources in the systems vary heavily with the arrival and departure of applications. The other is that the applications themselves, e.g., a multimedia application, may require different amount and type of resources over time. In the following, we first introduce the features of adaptive applications briefly, and then present the general methods and the key issues for using the adaptation technique.

2.3.3.1   Adaptive Applications

Some applications have little concern about the changes of the resources obtained from the network infrastructure. They do not work better if more resources are available to them. Other applications can adjust their behaviors according to the amount of resources they acquire, and their performance varies to some degree with the amount of resources consumed. The latter are normally called adaptive applications. Generally, many applications can be adaptive, including TCP-based applications, reliable multicast applications and many audio and video transmission applications. For example, some video applications can adjust their data transmission rate when they detect congestion in some network nodes.

The advantage of adaptive applications is that they can make full use of the network capabilities to obtain their best performance under different network conditions. Moreover, some network performance fluctuations can also be tolerated by the applications.

2.3.3.2   Adaptation Method

Basically, two tasks are involved when performing the resource adaptation, namely resource/performance monitoring and resource adaptation.

Resource/performance monitoring means that a mechanism for monitoring the application performance or the status of the available resources must exist, so that adaptations can be performed depending on the variance of the performance or the resource status. The resource/performance monitoring is the basis for the decision of the adaptation.

Resource adaptation means that applications adjust their traffic depending on the performance variance or the new resource conditions using some algorithms, or the networks adjust the resources allocated to the applications according to some criteria.

In general, where and how to perform the adaptation are the key issues for adaptation. According to the location where the adaptation is made, the adaptation can be divided into application-side adaptation and system-side adaptation.

Application-side adaptation

Application-side adaptation means that the mechanism for making adaptation decisions, i.e., when the adaptation will be triggered, and performing adaptations is embedded in applications. Projects such as [BS91] [HSN+97] [JLD+95] and [SM99], belong to this category. Since adaptations are performed according to the variance of application performance or resources allocated to applications in the network, some form of feedback loop between end-users and network nodes is needed, so that applications can acquire the network status information to adjust their performance.

In general, the application-side adaptation has the following features:

- Since the applications know the relationship between their performance and the resource requirement, they can adjust themselves directly according to the resource variation. Normally no translation between the performance and resource requirement is needed.

- Applications may be complex. Applications need to have a performance monitoring mechanism for processing the feedback information from the network, to decide when and how to perform adaptations.

- Application-side adaptation is relative slow. Applications cannot adjust their behavior promptly to follow the resource variation in the network, since they have to wait for the feedback from the network.

- The continuous feedback and negotiation information between applications and networks may result in the increase of the transmission overhead.

- The network lacks the knowledge of the global state necessary to preserve fairness among competing applications when the network resource status changes.

- The network has no means to access application level semantics; therefore, cannot take into account the individual adaptation capabilities and constraints of applications.

System-side adaptation

Adaptations can also be done at the system-side. That is to say, the network node system decides when and how to perform adaptations on applications. Work such as [HW96] [RS96] and [RSY98] adopts this method. Compared with the application-side adaptation, the system-side adaptation has the following advantages:

- The reaction to the resource status variations can be fast. Since the network nodes themselves posses all the information related to the decision of the adaptation, the adaptation can occur immediately after some resource changes are detected. This omits the time spent on information exchange between applications and network nodes.

- It eliminates the complex performance monitoring and control system in applications. In this case, applications do not need to care about the variance of their performance.

- It operates effectively as part of the integrated application-network system. Instead of requiring each application to hold an adaptation algorithm and mechanism for monitoring the performance or resource status in the network, one or several adaptation algorithms at the network nodes may serve a lot of applications.

However, the system-side adaptation needs to know about the relationship between the performance of an application and the corresponding resource requirement. This can be realized through two ways. One depends on the translation mechanism in applications. Namely applications inform the network in advance about the relationship between their performance and the corresponding resource requirement. During adaptations, the network does not need any further information from the applications. The other is that the network gets the information dynamically from the applications. The former is obviously very quick regarding adaptation time, and can also reduce the transmission overhead caused by the information exchange between applications and network. This method can best represent the advantages of the system-side adaptation. However, due to the complexity of the applications, it is difficult to find a general method to describe the relationship between the application performance and the resource requirement. Some research has considered models for describing the relationship in some specific application domain, such as multimedia [HW96] [NS95], or applications demanding a fixed set of resource and service types [AAS97]. The latter, however, causes some transmission overhead, just like the application-side adaptation. In addition, the network has to maintain the feedback loop information for each application that is adaptable. This increases the complexity and the cost of network node systems.

For both the application-side and the system-side adaptation, no matter which method is used, making adaptation decisions, i.e., under what conditions to trigger adaptations, is the key point. It determines the efficiency of the adaptation algorithm and the utilization of the whole system resources. Currently, the basis for making adaptation decisions includes the performance-based and the economics-based methods.

Performance-based resource adaptation

Most adaptation systems make adaptation decisions according to the variation of the performance of the applications. Namely the performance of the applications, usually measured by some QoS parameters, is monitored constantly. When the performance variation excesses a certain value, an adaptation is triggered. In recent years, in the context of continuous multimedia applications, substantial effort has been made to develop QoS-based resource adaptation components and systems, such as [CEA95], [SDL99] etc.

Economics-based approach to resource adaptation

Some research work advocates using economical models to encourage applications to compete for resources. Based on this theory, in some adaptation systems, the adaptation decisions are made according to the utility of the system, i.e., the economical gain that the system can earn by selling resources.

[KMT98] and [LL99] have proposed an algorithm which controls the flow rate by maximizing the aggregate source utility. Each source is characterized by a utility function U_s, which is an increasing and strictly concave function of source rate x_s. The objective of the network is to optimize the social welfare, in other words, to maximize the aggregate source utility, namely $max \sum_s U_s(x_s)$, subject to the capacity of each link, i.e., $\sum_s x_s \leq c_l$. In [KMT98], the objective is decomposed into optimization subproblems for the network and users. The user subproblem is to choose a willingness-to-pay w_s given the path price p^s in order to maximize its benefit, and the network subproblem is to choose source rates x_s given the willingness-to-pay of users in order to maximize $\sum_s w_s \log x_s$. It allows the users to decide their payments and receive what the network allocates. [LL99] solves the dual of the optimization problem. The basic idea is that each link in the network continuously adjusts its price and each source computes its optimal rate as a simple function of the total price of the links it uses. This approach allows users to decide their rates and pay what the network charges.

Based on the resource pricing mechanism [KMT98] [Clear96], [SM99] has proposed an adaptation architecture using frequently renegotiated timed resource contracts. The resource manager in the network is responsible for maximizing its revenue, which is generated by selling resource contracts to applications. Applications are responsible for maximizing their utility by purchasing and trading resource contracts. Applications are provided with credits from a User Agent, renewable over a given time-scale. User Agents are responsible for implementing the policy of a particular user of the system. In this architecture, the dynamic resource price is used as a congestion feedback mechanism which enables applications to make system policy controlled adaptation decisions.

[KL97] introduces a utility model for adaptive multimedia systems. In this model, each multimedia session has a quality profile, containing a set of operating qualities from the minimum acceptable quality to the maximum desired quality. The main task of the multi-session system is to find an operating quality for each session, in order to maximize the system utility. When a new session cannot increase the system utility, it will be rejected as unprofitable one.

2.3.4 Summary

In this section, we have presented a general overview about the resource management in the current IP-based network, focusing on two techniques providing QoS support for applications: resource reservation and adaptation. Generally, the resource reservation mechanism guarantees the resources used for realizing the requested QoS for applications. However, fixed-capability reservations tend to be wasteful of resources and hinder graceful degradation in the face of congestion. By using the resource adaptation method, resources allocated to applications vary dynamically. In this case, there is coordination between the application performance and the system resources, and normally the total system resources can be used well. Nevertheless, adaptation inevitably fail when congestion reduces available resources below acceptable limits[ABF97] [RLL+97].

2.4   Related Work

The last two sections have given an overview about ANs and resource management, concentrating on the key issues and general techniques covered in both fields. In this section, we discuss some work related to the AN node architectures and the resource management in active and programmable networks, which provides a close background of the work in this dissertation.

2.4.1   Active Network Node Architectures

As mentioned in section 2.2.2, the most fundamental components of an AN node architecture are NodeOS and EE. In this section, we classify the projects related to the AN node architecture in terms of EE and NodeOS, although some projects such as SwitchWare [AAH+98] and Janos [Janos] involve both EE and NodeOS.

2.4.1.1   Execution Environment (EE)

To date several EE architectures have been defined and implemented, including ANTS, PLAN, SNAP, ASP and CANE. ANTS [WGT98] [Wet99a] supports capsules that can carry both data and programs. It has an on-demand code distribution mechanism, through which the code of methods for processing capsules can be dynamically distributed to network nodes where they are needed. We will discuss ANTS in chapter 6 in more detail, since it has been used as the base of our implementation.

PLAN (Packet Language for Active Networks) [HKM+98] is based on the simply typed lambda calculus and provides a restricted set of primitives and data types to support basic data transport and layering of network protocols. PLAN is a part of the three-level active networking architecture SwitchWare and designed to be simple, safe and secure. PLAN itself is a strongly-typed functional language with syntax similar to the standard ML [MTH+97]. It supports standard programming features, such as functions and arithmetic. A notable restriction is that functions may not be recursive and there is no unbounded looping, which helps to guarantee that all PLAN programs will terminate. To support packet transmission, PLAN possesses primitives for remote evaluation: a user can specify that a computation (function call) should take place on a different node. Two prime primitives in PLAN are *OnNeighbor* and *OnRemote*. When used, both of them must be supplied with a resource bound acting as the TTL (Time To Live) or the hop counter. The resource bound is decremented on each hop; and during execution, a program must donate some of its resource bound to the *OnRemote* or *OnNeighbor* used for sending new packets. In addition, there has been further research to make PLAN more secure through trust management [HKS02]. PLAN provides the ability to manipulate programs as data, via a construct known as a chunk. Chunks provide the means for PLAN packets to be fragmented or encapsulated within one another. A major advantage of PLAN is that it is easy to use. Despite the limits on its computational power, it is a high-level language both in its syntax and its features. However, transmitting and receiving PLAN packets require marshalling and unmarshalling some representation of PLAN programs. This significantly increases the cost of PLAN packets compared with other conventional approaches. Moreover, the program termination guarantee is rather weak.

SNAP (Safe and Nimble Active Packet) [Moo02] is a second-generation active packet system designed to address the open problem of providing a flexible programming language with high performance, yet safe for execution. SNAP is a stack-based bytecode language, requiring no unmarshalling and permitting in-place execution. A packet contains code, heap, and stack segments. The stack is the last segment in a packet, allowing SNAP to be executed in-place in a network buffer, as the stack can grow and shrink into the available space at the end of the buffer. SNAP is indeed a low-level, assembly-style language, making tightly hand-tuned packet programs possible. A key aspect of the SNAP design is its model of predictable resource usage: SNAP programs use time, space and bandwidth linearly in the length of the program. This means also that SNAP programs are terminated. In the current implementation, SNAP programs are carried in IPv4 packets with the Router Alert option octet. Thus, the conventional routers will simply forward the SNAP packets toward their destination, whereas the SNAP-aware routers can detect the router alert, and check if a packet contains the SNAP program in the IP protocol field; if yes, the SNAP interpreter is called. In addition, each packet uses also a resource bound (RB) field for hop count.

ASP (Active Signalling Protocol) [BLB+02] is a Java-based EE designed to assist the development and deployment of the complex control-plane functionality such as signalling and network management. In essence, ASP offers a user-level operating system to active applications (AAs), by making use of the services provided by the underlying NodeOS and the Java programming language, and enriching some new functions, such as the enhancement to the network I/O construction. E.g., ASP uses the *Netiod* to replace the socket interface of the input and output channel abstraction in the general active node architecture. The interface between AAs and the ASP EE is called *protocol programming interface* (PPI), which allows AAs to access the resources in active nodes. Moreover, ASP has also defined an interface to the user applications (UA) in the end systems. The UA is considered to be "adjacent" to the ASP EE, it communicates with the ASP EE using a TCP connection. Currently, the ASP EE does not support in-band code loading, but rather supports an out-of-band loading mechanism. The ASP EE requires active packets to include an *AASpec*, which provides a reference for the code executed. The ASP implementation executes each AA in a distinct process, which may contain multiple threads defined using the *AspThread* Class, a class similar to the Java Thread Class but with some slight enhancements. Being Java-based, ASP provides a certain amount of security inherited from the Java language. It increases the protection through employing a customized Java Security Manager. However, there is no mechanism to protect specific parts of the AA from other parts of the AA.

CANE (Composable Active Network Elements) [BCZ98] has two goals: to support the development of active applications that require reasonable forwarding performance, and to provide a framework for the modular construction of services. The programming model used in the CANE EE consists of a fixed part and a variable part. The fixed part, i.e., the underlying program, represents the uniform processing applied on every packet. And the variable part, i.e., the injected program, represents the customized functionality exerted on the packets. The specific points in the underlying program where an injected program can be executed are called slots. Composition of services is achieved by selecting an available underlying program and then specifying a set of injected programs. The injected programs may be node-resident or loaded from a remote site. The CANE EE provides a variable-sharing mechanism for the injected programs to communicate with the underlying programs. The underlying programs declare variables that can be shared, allocate space, and export these variables to make them eligible for sharing. The injected programs can declare references to shared variables. The instantiation of an underlying program and a set of injected programs at an active node occurs via the CANE signaling protocol. The signaling protocol sends a CANE user interface (CUI) message to the node where a new application should be supported. When a CUI message is received at a node supporting CANE, the underlying and the injected program code is fetched and a new application can be composed by these programs.

2.4.1.2 Node Operating System (NodeOS)

Various NodeOS have been designed and implemented by different research groups, including Scout, Janos, Snow, AMP etc. which are stand-alone NodeOS, and Bowman, PromethOS which are extensions of legacy OS.

Bowman [MBZ+00] runs on the generic UNIX substrate. Since it was developed in the early days of the NodeOS specification [AN01], it provides only a subset of the interface suggested in the specification. Janos [THL01] is a Java-oriented NodeOS. It has two primary research goals: resource management and control, and first class support for untrusted active applications written in Java. We will discuss this work further in chapter 6, as it has been used as the base of our implementation of the AN node system.

Scout [PGH+01] is a configurable system, where an instance of Scout is constructed by a set of modules. For example, a portion of the module graph for an active router may consist of 6 modules: module TCP, UDP, IP and ANEP, each implementing a communication protocol; module JVM and NodeOS, each implementing an API. Scout paths support data flows through the module graph between any pair of devices. When configured to

implement a router, Scout supports network-to-network paths, which is called forwarding path. The entity that creates a forwarding path specifies three pieces of information:
the sequence of modules defining how the path processes messages, a demultiplexing
key identifying what packets will be processed by the path, and the resource limits placed
on the path, including how many packets can be buffered in its input queue, the rate at
which it is allowed to consume CPU cycles, and the share of the link bandwidth that the
path may use. The same information is required by the NodeOS: a domain is a container
for the necessary resources (channels, threads, and memory), while a channel is specified
by giving the desired processing modules and demultiplexing keys.

The goal of AMP [DPS02] is to provide a secure platform upon which EEs and active
applications can run, without unduly compromising efficiency. AMP consists of library
code (LibAMP) and four trusted servers, which provide jointly the NodeOS interface to
an EE. LibAMP follows the exokernel library OS design approach of placing a copy of
the OS in the same address space as the application. To provide protection of system-
wide state information, portions of the NodeOS abstractions are implemented within
separate trusted servers, and LibAMP invokes protected operations via cross-address
space RPC. AMP is able to accommodate a wide range of EE implementations and
languages. Because there is relatively little trust that must be placed in a given EE, AMP
can limit the resources and operations that an EE is granted to access, thereby allowing
more flexibility in configuring EEs to run within the system.

Snow [SGW+02] is a NodeOS within the Linux kernel based on Silk, which encapsulated the Scout path architecture into a loadable Linux module. The design goal of Snow
is to develop a NodeOS that can be embedded in a wide-spread, open source operating
system. Snow allows non-active applications and regular operating system operations
to proceed in a regular manner, being unhindered by the active networking component.
It offers performance competitive with that of networking stacks of general-purpose
operating systems.

PromethOS [KRG+02] is a modular router architecture based on Linux kernel 2.4
which can be dynamically extended by plugin modules installed in the networking kernel, thereby the Linux-resident simple module management functions can be extended
for use in an AN environment. The plugin framework manages all loadable plugins
and dispatches incoming packets to plugins according to matching filters. Once a packet
arrives and needs to be processed by a plugin, the framework invokes the previously
registered plugin-specific callback function. A signaling protocol is used to retrieve plugins from remote code servers, install and configure them and setup network wide paths
to send them to the desired nodes. PromethOS also offers the demultiplexing functions
typically needed in an AN node, and is capable of linking kernel-space processing with
user-space higher level processes.

2.4.1.3 Summary

Till now, some architectures concerning EE and NodeOS have been implemented. Each of them has its own features. E.g., PLAN and SNAP are language-based EEs, ASP is a control-plane oriented EE, CANE stresses the service composition, Scout is a configurable system, Snow and PromethOS aim to implement a NodeOS on a widely used, open source operating system. In conclusion, the research issues covered by these systems focus on the following four aspects:

1. Extent of programmability. There is a debate on the extent to which the end user should be allowed to program an active node. The greater the level of programmability provided for the end user, the greater is the flexibility of an AN, nevertheless, the greater is the level of security risk or threat to the node stability. ANTS is one of the first AN implementations and conforms to the first approach of providing greater variety and degree of programmability for users. CANE is an example of the second approach in which the user is provided only a limited choice of services from which to choose from.

2. Processing performance. Since ANs involve user-directed processing of data, i.e., the active nodes have to interpret the programs or instructions carried in the packets from the user applications and then change their mode of processing to support the user-demanded applications, much processing is needed in AN nodes. It is argued that the AN technologies should be able to provide comparable speed compared with the traditional passive nodes in addition to providing programmability. SNAP has addressed this issue. In addition, the PAN architecture concentrates also on this issue.

3. Security issues. ANs can be threatened in terms of (i) the corruption of the NodeOS, EE(s) or the codes from users cached at the node by user applications intentionally or unintentionally; (ii) the consumption of large amounts of node resources by a single or few processes leading to failure of other services or processes; (iii) the manipulation of variables of a user process by another user process through the incapacity of the isolation mechanism provided by the NodeOS. Therefore, providing greater security to ANs and the AN nodes has been the focus of attention for many systems, such as SNAP, AMP, Janos, and RCANE.

4. Compatibility with legacy systems. In order to provide for the deployment of ANs, on the one side, active nodes should be able to interact with the traditional network nodes, which has been considered by most of the work. On the other side, the active node architecture should also be easily integrated into the widely used systems. Snow and PromethOS have addressed this issue.

2.4.2   Resource Management in Active and Programmable Networks

2.4.2.1   Resource Management in Active Networks

Due to the importance of the resources for active nodes, research work related to resource management came up with the birth of ANs. Resource management first attracted attention in ANs for the sake of security. The mobile code should not jeopardize the active node system and other active applications through overusing or exhausting the system resources. Therefore, some measures have been taken against the abuse of system resources. ANTS [WGT98] [Wet99a] uses a TTL approach to limit the times that a capsule can access the active nodes; PLAN and SNAP have both inherited this idea.

SNAP [HMN01] adopts the subtly designed language to count and limit the resources used by applications. All the SNAP instructions are designed to execute in constant time and space, and all the branches must move forward. Therefore, the SNAP programs consume time and space proportional to the packet length. In other words, if the SNAP programmers can express their programs in SNAP, those programs are resource safe. Correspondingly, the SNAP routers are sure that the incoming SNAP programs will behave reasonably concerning the resources even without examining them. In addition, each packet has an associated resource bound (RB) field that is decremented at each network hop, including sending packets to the current node via a loop back interface. Moreover, a packet must donate some of its RB to any child packets it spawns. As a result, a strict upper bound on the amount of network resources a packet or its descendants ultimately use can be easily placed. The primary goal of the resource controlled active node environment (RCANE) [Men99] is to resist many classes of denial of service (DoS) attacks. RCANE supports an AN programming model over the Nemesis Operating System [LMB+96]. It is designed to provide resource isolation between multiple independent applications in a programmable network node, supporting robust control and accounting of system resources, including CPU and I/O scheduling, and garbage collection overhead. RCANE uses the abstraction of a session to represent a principal requiring resources in the node. A session is similar to the concept of a domain in the NodeOS API [AN01]. Sessions are isolated, so that activities occurring in one session cannot prevent other sessions from receiving the guaranteed resources, except in situations where explicit interaction is requested (e.g., due to one session using services provided by another session). The use of sessions in RCANE makes the end-user the resource principal, allowing guarantees to be provided more easily to individual end-users. An EE then becomes a library that a session may use to provide a convenient programming abstraction, and a client may make use of more than one EE in a single session if desired.

RCANE's CPU management abstractions, such as the Virtual Processor (VP), thread, and thread pool, are structured so as to allow sessions to split their tasks between multiple scheduling classes, control the level of concurrency used for different sets of tasks, and to serve those tasks efficiently. E.g., a session may have one or more VPs. By requesting multiple VPs, a session may organize its tasks into multiple independently scheduled classes. In addition, each thread pool is associated with a particular VP. Its threads are only eligible to run when its VP receives CPU time. RCANE allows sessions to open channels to access network resources. A channel may have a guaranteed level of buffering and transmission bandwidth. Each channel capable of receiving packets is associated with a demultiplexing specification and a VP, which is used for receiving protocol processing on that channel. To prevent the crosstalk between the network activities of different clients, all packets are demultiplexed to their receiving pools using a packet filter. The RCANE memory architecture is based around the model of multiple garbage collected heaps in a single virtual machine. Each session has its own independent heap. An incremental garbage collector, which processes a small portion of the heap each time it is invoked, rather than processing the entire heap in a single invocation, is employed to prevent excessive interruptions to execution. Such a property is essential to prevent the unpredictable nature of garbage collection from causing clients to miss their deadlines.

The research group in NIST concentrates on predicting and controlling resource usage in a heterogeneous AN [GMC01a] [GMC01b]. Their starting point is that the resource requirement, especially the CPU requirement, from the mobile codes is different at different network nodes, which have different levels of CPU speed, memory capacity as well as OS etc. Hence a model is needed to predict the CPU resource requirement in the intermediate nodes according to that in the source node. They use AVNMP (Active Virtual Network Management Prediction) as an overlay network, over which a simulation model runs in advance to predict the resource demand in each intermediate AN node.

[AHI+00] presents a market-based resource management mechanism implemented on the ALIEN model [Ale98] of the SwitchWare project. The physical resources of an active node, e.g., CPU, memory, network capacity and secondary storage, are regarded as the commodities traded in the AN economy. The producing agents are the elements of an active node, and the consuming agents are active programs that wish to run on an active node through using the elements of the node. An agent of the service broker peddles resource access rights, which are essentially policy rules governing the use of AN resources, e.g., what function may be called, when it may be called and if there are any restrictions on its arguments. The KeyNote trust management approach [BFI+99] is used to generate the relevant policies and delegation credentials, as well as to specify the local policies and credentials. In this system, resource access rights are implemented as polices initially authorized by the resource producer. At the outset, these policies are applicable to the service brokers, who may then delegate them to the consumers who purchase them. Consumers then provide the policy credentials to the producer when they want to access the service. During the implementation, the Active Loader in ALIEN has been modified to control the visible API of active extensions in a more fine-grained manner, and some of the functionality of the resource schedulers has been exposed into the core functionality, so that they can be similarly controlled by the Active Loader. In addition, a service broker communication module (BCM) is implemented, whose task is to relay communication between brokers and users.

In addition to the work about resource management in active nodes, [SJ03] has introduced a framework for resource management in the whole AN. The goal of this work is to allocate and manage node resources in an efficient way while ensuring the effective utilization of the network and supporting load balancing. The framework supports co-existence of active and non-active nodes and proposes a Directory Service (DS) architecture which can be used to discover the suitable active nodes in the Internet, select the best network path (end-to-end) and reserve the resources along the selected path. In addition, an active node database system and a simple adaptive prediction technique have been used to determine the CPU requirements of the incoming packets.

2.4.2.2   Resource Management in Open Programmable Networks

Beside the work in the field of ANs, Spawning Networks [CKV+99], which belongs to the community of the open programmable interface networks, has implemented a resource management system, the Virtuosity, in its Genesis Kernel [CVV99] [VCV01]. The spawning networks support the dynamic creation and deployment, i.e., spawning of virtual network architectures through profiling, spawning and management procedures.

The Virtuosity is responsible for the resource management of the spawned virtual networks, such as allocating virtual links to child virtual networks or customers. Since the Genesis Kernel creates a natural hierarchy through partitioning and isolating the virtual networks, the Virtuosity framework is designed to have the functionality of inheritance, autonomous control and slow time-scale dynamic provisioning of network resources. It comprises a number of distributed elements, which are initiated as part of the child virtual network kernel during the spawning phase and deployed as distributed objects. They operate within the child and parent virtual network kernels, managing the partitioned resource space and interfacing with the parent virtuosity system to increase or decrease the current partitioned resource space through dynamic provisioning. The key component, the maestro, coordinates virtual network control through a set of distributed virtuosity components performing virtual network monitoring, renegotiation-based resource allocation and capacity-based scheduling, all of which operate over management-level timescales. The maestro also influences the way in which resources are allocated to its child networks by optionally setting market pricing strategies or alternative resource allocation strategies. During the spawning phase of a child network, the maestro conducts a virtual network admission control test based on the resources requested by a child network. If the admission control test is positive, then the parent provider network admits the child network and allows it to become a participant in the resource allocation process governed by the policy of the child virtual network.

Another key component of Virtuosity is its capacity scheduling capability. A virtual network scheduling abstraction resident at each parent resource arbitrates child virtual network access to the parent virtual links. The capacity arbitrator is based on a set of virtual network capacity classes and capacity class weight policies that are distributed to the arbitrator by delegates on behalf of the maestro. Capacity classes represent virtual network differentiated policy for provisioning capacity. The capacity classes and weights translate the resource allocation negotiated by individual child virtual resources during the resource negotiation process to a set of virtual capacity scheduling policies. These policies are then used to differentiate child virtual network capacity allocations and the ordering of packet delivery to the parent link resource.

The core router operating system support (CROSS) project [YC01] is concerned with the development of an OS that will be used in software programmable routers, which can provide value-added services, such as security, accounting, caching, and resource management. CROSS concurrently supports multiple virtual machines, each of which is able to obtain a guaranteed share of physical router resources. And each virtual machine can provide an independent service for the traversing flows (e.g., each ISP can deploy its own virtual machine to the routers). It employs hierarchical scheduling techniques that can provide guaranteed minimum share resource partitioning based on customizable system policies. CROSS defines a kernel abstraction called *resource allocation*, which is orthogonal to OS resource consumers like threads and process address spaces. Resource allocations can be flexibly bound to resource consumers at run time. They provide access to various time and space shared resources with QoS (such as throughput, delay, and proportional sharing) guarantees, each of them controlled by its own resource specific scheduler. Moreover, a packet embedding a CROSS API call must be channeled to its destination virtual machine. The embedded call must often run with dynamically bound resource allocations.

Resource allocations in CROSS are initially bound to resource consumers. This allows even unmodified "QoS-unaware" applications to run with definite resource shares. In addition, CROSS provides a new API for QoS-aware programs to manage these allocations. A main objective of resource allocations is to give CROSS programs fine-grained control over how they will receive system resources, in order to enable QoS guarantees and differentiations according to application requirements, user priorities, price payments etc. Furthermore, scheduling algorithms are designed in a resource specific manner.

2.4.2.3 Summary

Resource management has been emphasized by some active node architectures from the perspective of security, such as Janos, RCANE, and SNAP. They emphasize on limiting and isolating the resources that an application can use. Although RCANE allows sessions to reserve resources in advance, it focuses on isolating sessions so that activities occurring in one session cannot prevent other sessions from receiving their guaranteed QoS.

In addition, some further resource management methods have also been suggested in the active and open programmable networks. A market-based resource management method has been introduced in ALIEN, where resource access rights are sold by active nodes, and applications buy them using policy credentials. Moreover, the Virtuosity framework suggests a measurement-based admission control method with renegotiation mechanism to allocate resources to the child virtual networks. However, it focuses only on the virtual link resources.

2.5   Existing Work versus Our Focus

From the discussions in this chapter we can see that the AN node architecture is the core of ANs. It has been the focus of attention since the emergence of ANs. Much effort has been made on how to provide a flexible, yet efficient and secure AN node architecture, solving problems related to the programmability, security and performance. Work related to resource management in the AN nodes so far concentrates mainly on the security issues, such as isolating the user processes and the resources they use, controlling the resources used by the user program in order to prevent them from overusing or exhausting the node system resources. However, the provisioning of resource guarantees for applications in AN nodes has not been addressed sufficiently as done in the traditional IP-based networks. Especially, new emerged problems related to the resource usage in ANs, such as the possible relationship among different types of resources, the balance of the use of different types of resources in an AN node system, as well as the flexibility of the resource requirement of active applications, e.g., due to the use of multiple types of resources, different resource combinations may satisfy the need of the application, have not been considered and emphasized. Till now, no AN node architecture has been implemented that supports the provisioning of resource guarantee for applications and takes the above problems into account. Although some work in the field of programmable networks has involved providing resource guarantee for applications or the use of multiple types of resources, the applications can only be on the management level, and each resource type has been treated separately without consideration of the features of applications.

On the other side, resource management has been widely discussed in the traditional IP-based networks. Some resource management mechanisms, such as reservation and adaptation, have been presented. Some resource management models and architectures, such as Intserv and Diffserv, have been established and implemented. However, the resource management methods used in the traditional networks concentrate mainly on the network bandwidth resources. The new resource usage characteristics emerged in ANs, such as the demand for multiple types of resources, and the flexibility of active applications, result in that these methods cannot be directly used in ANs.

Hence, the new emerged problems related to the resource usage in ANs should be studied, such as the use of multiple types of resources and the relationship among different types of resources, the flexibility of applications, as well as the balance of the system resource usage. New mechanisms are needed in order to provide resource guarantees for applications in AN nodes and at the same time take these problems into account, satisfying the need of applications and making the limited resources in the AN node systems well utilized. This motivates the work in the dissertation, which suggests a new resource description method in ANs and presents an adaptive resource management mechanism for the AN nodes to emphasize and solve these problems, and implements an AN node architecture supporting this mechanism.

2.6 Summary

In this chapter, we have presented some background information about ANs, including the general node architecture, the resources needed in the node system, the active applications and the active network encapsulation protocol. We have also given a general overview about the resource management approaches currently used in the IP-based networks. In addition, related work has also been discussed in the context of AN node architecture and resource management in the active and programmable networks, concentrating on their features and aspects that are influential on the ideas of our work. These provide a necessary basis for the remainder of this dissertation, which presents an adaptive resource management mechanism, implements an AN node architecture supporting this mechanism by selecting ANTS and Janos as the EE and NodeOS respectively.

As a summary, we emphasize the following issues:

First, ANs consist of a number of network nodes, e.g., routers, which possess mechanisms supporting the loading and execution of user-specified programs. These nodes are called AN nodes or active nodes. In addition, an AN can be considered as an overlay network upon the traditional IP-based network, namely the active packets can pass through the traditional routers transparently.

Second, programs, also written as user-specified or application-specified programs, mean a collection of instructions written in a platform independent language that can be carried in active packets, and be loaded into and executed in the active nodes.

Chapter 3

Resource Vector

Last chapter has introduced the basic architecture of AN nodes and presented the abstractions concerning the resource usage in the NodeOS suggested by the AN research group.

This chapter first analyzes the resource requirement and consumption in AN nodes from the perspective of active applications, and summarizes the characteristics of the resource usage. Based on this, the challenges to the resource description and management in AN nodes are put forward. Then a method, called resource vector (RV), is suggested, which reflects the resource usage characteristics and can be used to describe the resource requirements of active applications and their resource usage in AN nodes. Following this, the operations and features of RV are presented. And finally, the idealized system resource consumption model based on RV in an AN node is introduced.

3.1   Resources Required by an Active Application

ANs are distinguished from the traditional IP-based networks because of the introduction of the application-specific computation into the network nodes. Packets from the active applications are allowed to carry and specify programs, which are executed at the network nodes processing these packets. However, the execution of the application-specific programs needs support of more types and amount of resources in the network node system compared with the traditional networks, where typically only data forwarding is performed in the network nodes. More types and amount of resources need to be allocated to an application. Basically, these resources are used for packet receiving, processing and sending. In the following, we discuss the various resources needed by an application at an AN node in detail through comparison with that in the traditional networks.

3.1.1   CPU Cycles

In the traditional IP-based networks, normally the CPU cycles cost by an application is roughly proportional to the bandwidth offered to the application[1]. Packet processing means routing and forwarding. Network nodes analyze the header of the packets, looking up the outgoing line to use for it in the routing tables and forward them to the corresponding next hop or destination. The CPU cycles consumed for routing are approximately constant per-packet, thus the total CPU cycles used for routing is related to the total number of packets passing through the node. Packet sending and receiving has either hardware support or implies copying between network devices and node system, in which case the CPU cycles cost is basically proportional to the incoming bandwidth. Hence, in the traditional IP-based networks, normally only the network bandwidth resource is stressed[2,3].

However, in AN, the packet processing becomes very complicated. Besides routing, CPU cycles are also needed for the execution of the user-programs, which are typically a series of instructions written in platform independent languages.

Figure 3.1 illustrates an example of the user-specific program. Typically the program can be divided into two categories. One is the system function or service calls, i.e., the NodeOS has already provided some basic functions or services for the applications from the end-users, which can be invoked by applications through the interfaces in their execution environments. The other is code written by the developer of the application, say application code. Correspondingly, the CPU cycles used for processing the programs from the end-users include two parts: one spent on executing the application code, and the other spent on executing the system function calls. The amount of the former varies with the application, and the latter includes also tasks like garbage collection etc. Moreover, the application code is also quite flexible. They can even decide what to do further according to the system status or some results of the program execution, for example, if new packets or new processes should be generated or spawned at this node.

In addition, if applications use the integrated-code approach to carry the programs, the CPU cycles used for loading the program into the execution environment might also not be neglected.

On the other side, the CPU cycles spent on packet receiving and sending cannot be simply regarded as proportional to the incoming network bandwidth anymore. They are related also to the user-specified programs, because new packets can be generated at a network node due to the execution of the user-specific programs. Moreover, because of the complexity of the resource organization and classification in AN nodes, sending and receiving new packets does not mean only copying, e.g., new InChannel and OutChannel

---

[1]This is not the case if control messages of e.g., transport protocols, and some of the higher level functions of IP, such as subnet broadcast, are taken into account.

[2]The CPU cost for some control purpose, such as congestion control or QoS provisioning, is usually considered through separate ways.

[3]For different types of applications, the average packet length is not the same. However, this does not prevent from estimating the CPU usage using bandwidth.

```
Example() {
    If (thisNode==concernedNode)
        bandwidth=getBandwidthInfo();        //this function is provided by the NodeOS
        if ( bandwidth>100 )
            generateNewPacket(100);          //this function is defined by the user itself
        else if (bandwidth >50)
            generateNewPacket(100);
        else goAhead();
    else
        goAhead();                           //this is also a user defined function
}
generateNewPacket( int num) {
    for (int i =0; i<num; i++) {
        cap=new Capsule();
        thisNode.outChannel.send(cap);
    }
}
goAhead() {
    thisNode.outChannel.send(this);
}
```

Figure 3.1: An example of user-specific program

need to be created. Hence, this part of the CPU cycles might also not be neglected.

3.1.2 Memory

In the traditional networks, during the packet receiving, processing and sending, memory is used to store the packets, i.e., for buffering. The size of the needed memory is tightly related to the number of the packets and their size. However, in AN, besides buffering, memory is also needed for the loading, execution and caching of the application-specific programs. The concrete amount depends normally on the platform of the network nodes. Generally, the memory usage related to a user-specified application falls into five categories:

1. Packet buffers: packet buffers provide a level of queuing for both transmission and reception at network interfaces. Just like in the traditional IP-based networks, in AN, a client cannot be guaranteed exclusive access to the CPU, and the latency in the paths between the network device drivers and the client is likely also not negligible. Therefore, packet buffers are still necessary.

2. Code modules: since application-specified code is allowed to be injected into the network nodes, space must be provided to store the code. Besides the instructions, some form of persistent variables may also be included in the code modules. How long the code will be stored in the node, who and how many users can access these code modules dynamically loaded, depend on the node system implementation.

3. Thread stacks: Thread stacks are the context associated with a computation occurring on the node. Various events, such as processing or forwarding a packet, or waiting for another packet etc., may occur during the computation. Thread stacks represents a measure of concurrency, which can help to improve the total processing rate and response time. Temporary variables and partial results are normally stored in the corresponding stacks.

4. Heap memory: in some operating systems, such as Java Virtual Machine [LY99], the heap is a piece of memory shared by some threads having some common characteristics, e.g., the threads belonging to the same Java class. In an AN node, a heap is typically used to share some persistent data among different flows or among the packets belonging to the same flow. Examples include the timed soft-states in PLAN [HMA+99], where data may be flushed from storage by the node if the space is needed for other purposes. And the code cache in ANTS [WGT98], where the Java classes for a particular type of packets are cached along the path taken by those packets.

5. Memory for other data: In order to maintain the normal operation of the node system, the system allocates a data structure to manage all the stacks used by each incoming application. In addition, memory is also needed to store the accounting and billing information of each application.

3.1.3   Network Bandwidth

Network bandwidth is needed for packet sending and receiving. In the traditional networks, the bandwidth offered to a flow is the fundamental metric of quality of service. Other types of resources that the flow needs, such as the CPU cycles and memory, are correlated with the incoming bandwidth the flow consumes. Moreover, the outgoing-bandwidth, namely bandwidth for sending packets from a node, is only related to the number of packets and their size at the incoming side.

In AN, since the primary function of AN is still communication and not computation [Cal99], the main resource demand of applications in AN is still bandwidth. However, unlike in the traditional networks, the outgoing-bandwidth in AN is also related to the application-specific program and must be emphasized. Since new packets may be generated due to the execution of the programs.

3.1.4   Other resources

Other resources needed in AN include the code backup/caching store, persistent store, specialized hardware for encryption, special/leased links, as well as some special data, such as routing table entries. However, they are not the focus of our discussion, since they are not present at all AN nodes. We consider only CPU cycles, memory and network bandwidth resources in this work.

3.1.5 Summary

In short, in the traditional IP-based networks, CPU resource is needed mainly for routing and copying, and memory is used to buffer the packets temporarily. Both of them are correlated with the bandwidth that an application requires. Hence, typically only network bandwidth is discussed in the traditional IP-based networks. Furthermore, it is relative easy for end-users to estimate the bandwidth resources they need, and for the network nodes to monitor or calculate how many bandwidth a user has consumed.

However, in AN, the processing at the network nodes is not limited to a well-defined, predetermined function such as routing, scheduling and queuing etc. Applications can spawn processes at the AN nodes and may perform functions such as compression, duplication, caching etc. The straightforward relationship between the network bandwidth and other types of resources does not exist any more. Both the computation and the memory resource consumed by a user application have to be considered separately.

3.2 Characteristics of Resource Consumption in AN Nodes

Resource usage at the AN nodes has changed greatly due to the introduction of the application-specific computation in the network nodes. Different types of resources need to be taken into account. Moreover, due to the flexibility of applications, the consumption of one type of resource may interact with other types. In other words, some interior relationship among different types of resources may exist from the perspective of applications. In the following we summarize the characteristics of the resource usage at AN nodes, and analyze the possible influence of these characteristics on the resource management in AN nodes.

3.2.1 Resource Consumption Characteristics

In general, the characteristics concerning the resource consumption of applications at the AN nodes can be summarized as follows:

Multi-dimensionality: as addressed in the above sections, multiple types of resources are needed to execute the application-specified programs at AN nodes.

Complementarity: the needs of different kinds of resources by an application is not fixed, furthermore, the needed amount of a certain type of resource can sometimes interact with another. For example, when there is not enough transmission bandwidth while sufficient computation resource in a network node, a videoconference application may consume more CPU resource to compress the audio and video data more highly, such that the requirement for network bandwidth for transferring data decreases. Through this method, the ultimate performance of the application can still satisfy the user. That means the CPU and the network bandwidth resource can be complementary to each other. Another example is the representation of certain data structures, like sparse matrices and hashtables, where the well-known tradeoffs exist between storage space and element access time. Caching can also be considered as an instance of complementarity between memory and network bandwidth and/or CPU time used for information retrieval.

Higher-level sharing: in the traditional IP-based networks, usually network resource is shared among links or connections. But in ANs, in order to execute an application program safely at a network node, an execution environment (EE) is allocated to it. Resources at a network node system are allocated to different EEs and can only be consumed by applications through EEs. I.e., resources at an AN node are shared among EEs. In fact, an EE can be regarded as a virtual machine, which maintains different types of resources simultaneously, and in which a user application is executed.

3.2.2   Influence on Resource Management

The above characteristics put forward challenges to the resource management in AN nodes in the following aspects:

- Resource description: considering the multi-dimensionality and complementarity, resources allocated to an application in ANs can in fact be variable. Therefore, a simple list of different kinds of resources such as the bandwidth, CPU cycles and memory size, as used in the current projects involving multiple types of resources like [Men99] and [YC01], cannot express the resource requirement by an application well. Hence, a new resource description method concerning the multi-dimensionality and complementarity, and simultaneously reflecting the high-level sharing characteristic is necessary.

- Resource admission control: in order to provide performance guarantee for applications, before admitting an active application into an AN node, not only the transmission bandwidth, but also the processing and memory resource at the AN node must be checked.

- Resource allocation: in AN nodes, transmission, processing and memory resource are allocated to the EEs on behalf of the active applications. The challenge here is due to the complementarity among the different types of resources. Allocation of more CPU resource may reduce the amount of transmission bandwidth required on the outgoing link. Hence, a good resource allocation algorithm should be able to take the system resource conditions into account and make a compromise between the available resources in the node system and the resource requirement of various applications, in order to make full use of the system resources.

Resource description is the basis and influences the resource admission control and allocation algorithm. Only with a proper resource description scheme, the amount of the needed and allocated resource can be expressed.

3.3 Resource Description

3.3.1 System Resource Organization

As suggested in [Cal99], functionally an AN node consists of two parts: execution environments (EEs) and the node operating system (NodeOS). EEs are in fact virtual machines in which user processes are executed. They are the "sterile" environments provided by the node to execute each user-specified program. In addition, there may be multiple types of EEs in one AN node, applications can specify in their packets to the networks, which type of EE they use. The NodeOS multiplexes the communication, memory and computational resource among the various packet flows traversing the node, and provides basic functions that can be invoked by EEs.

Domain is the primary abstraction for resource accounting, admission control and scheduling in the system, as mentioned in the last chapter. A given domain is created in the context of an existing domain. The NodeOS is corresponding to the root domain, whereas the EEs are at the second level. EEs create domains to process packets from users. Since there may be multiple types of EEs in a node, and the EE implementation mechanisms may be different, the ways and methods that EEs create domains to process the packet flows are different. Some EE implementations may choose to process independent packet flows in their own domains, namely a domain may be created for processing a packet flow, such as the EE type 1 in Figure 3.2. In this case, usually the system creates a domain (i.e., an EE) for executing one specific program, also called protocol, and all packet flows requiring this program are first passed to and processed in this domain. We call this domain the protocol-level domain. And then depending on the program and each packet flow, a new domain may be created in the context of this protocol domain to process the packet flow continuously. We call this domain the packet flow or application-level domain. Other EE implementations may aggregate all packets on a single set of channel, memory and thread resources, such as the EE type N in Figure 3.2. In this case, all packet flows requiring the same program are processed in one EE, sharing the resources owned by this EE. In general, both the protocol-level and the application-level domains are in fact environments created on behalf of user applications. Therefore, we call them EE instances. An EE instance encapsulates the bandwidth, memory and CPU resource needed for executing an application-specific program, and is the entity for resource admission control, accounting, scheduling in the node system.

Therefore, we use a hierarchical model to organize the system resources, as shown

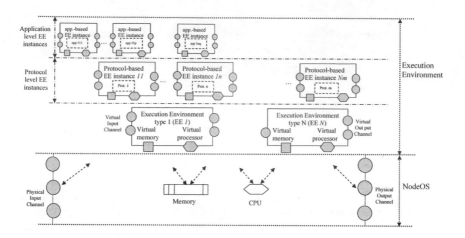

Figure 3.2: Resource organization in AN nodes

in Figure 3.2. The system resources are assigned to different types of EEs[4]. Each type of EE possesses a certain amount of the system resources. The resources owned by each type of EE are dynamically allocated to different EE instances created on behalf of active applications based on RVs introduced in the next section. In the following, we assume that an EE instance is created when a new application arrives at a node system, which means implicitly either an application level EE instance will be created for processing the packet flow of this application, or the protocol level EE instance for processing this packet has not existed, and a new one must be created.

3.3.2 Resource Vector (RV)

As mentioned above, an AN node may support multiple types of EE. When a request of a new application comes, an EE instance of the specified type is created for the application and a certain amount of resources is allocated to this EE. Hence, our system has the following parameters:

- l: the number of EE types in a node architecture. For instance, if ANTS and PLAN can be simultaneously supported in an AN node, l is equal to 2.

- $m^j (1 \leq j \leq l)$: the number of the EE instances existing in the jth EE type at one time point. Therefore, the total number of instances in the system is $m = \sum_{j=1}^{l} m^j$. Note that the EE instances can be at the protocol level or at the application level, which can be configured at the system initialization.

---

[4]The resource amount assigned to each EE type can be determined according to statistics of the number of applications supported by each type of EE and their function.

- n: the number of the resource types considered in a node system. For example, if only CPU, memory and bandwidth resource are of concern within a node, n is equal to 3.

Note that the total amount of resources owned by each type of EE is dynamic in "long-term". At the system initialization, a certain amount of resources is assigned to each type of EE, based on statistics of its resource demand, e.g., the statistical number of applications that it has to serve in a definite time period, as well as the resource requirement intensity of those applications. An application in one type of EE borrows or lends resources from/to applications in other types of EE only in the case of emergency.

In the following discussion, subscript denotes variables related to different instances, whereas superscript represents variables related to the type of EE. Boldface stands for a vector. E.g., $\mathbf{r}_i^j$ represents the resources used by the ith instance in the jth EE.

3.3.2.1 RV Definition

Let $r_{i,j}$ denote the amount of the jth type of resource used by the ith EE instance I_i, then the total resources occupied by I_i can be presented by $\mathbf{r}_i = \{r_{i,1}, r_{i,2}, \ldots, r_{i,n}\}$. We call $\mathbf{r}_i$ a resource vector (RV) of I_i.

Based on the concept of RV, the resources in an AN node system can be imaged as a n-dimensional resource space, with each resource type spawning a dimension. Thus, the resource $\mathbf{r}_i$ used by the EE instance I_i is a point in the resource space and $r_{i,j}$, one type of the resources required to guarantee the corresponding application performance of I_i, becomes a coordinate of $\mathbf{r}_i$ in the resource space. From another point of view, when a point in the resource space is determined, the corresponding resources, including their types and amount, can be determined accordingly. Furthermore, multiple points with some commonality in the resource space can present the complementarity: one point can be replaced by another one. Moreover, a part of the resource space can also be used to describe the resource requirement of an application: all points in this part of space can satisfy the users of the application.

3.3.2.2 RV Operations

Using this vector assumption, vector operations can be directly applied to RVs, such as plus, minus, dot product and equal, just like other general vectors. In order to be able to use RV directly for resource management, such as resource admission control, adaptation and allocation, we have also extended some operations to RVs, e.g., division and comparison. The operations that can be applied to RVs are summarized as follows:

Suppose $\mathbf{r}_1 = \{r_{1,1}, r_{1,2}, \ldots, r_{1,n}\}$, $\mathbf{r}_2 = \{r_{2,1}, r_{2,2}, \ldots, r_{2,n}\}$, then we have:

1. Plus: $\mathbf{r}_1 + \mathbf{r}_2 = \{r_{1,1} + r_{2,1},\ r_{1,2} + r_{2,2},\ \ldots,\ r_{1,n} + r_{2,n}\}$

   That is to say, the sum of two RVs is obtained through the sum of the two corresponding coordinates. The result is still a RV.

2. Minus: $\mathbf{r}_1 - \mathbf{r}_2 = \{r_{1,1} - r_{2,1}, r_{1,2} - r_{2,2}, \ldots, r_{1,n} - r_{2,n}\}$

   Similar to the plus operation, the difference between two RVs means the difference between the two corresponding coordinates. The result is still a RV.

3. Division: $\mathbf{r}_1 / \mathbf{r}_2 = \{r_{1,1}/r_{2,1}, r_{1,2}/r_{2,2}, \ldots, r_{1,n}/r_{2,n}\}$

   The division between two RVs is achieved through the division between the corresponding coordinates. Note that here the coordinates of $\mathbf{r}_2$ cannot be zero. The result of division is still a RV.

4. Compare, namely

   $\mathbf{r}_1 > \mathbf{r}_2$, if and only if $r_{1,j} > r_{2,j}$ for all j;
   $\mathbf{r}_1 < \mathbf{r}_2$, if and only if $r_{1,j} < r_{2,j}$ for all j;
   $\mathbf{r}_1 = \mathbf{r}_2$, if and only if $r_{1,j} = r_{2,j}$ for all j;

   Note that the comparisons have only partial orders.

5. Dot product: $\mathbf{r}_1 \bullet \mathbf{r}_2 = r_{1,1} \times r_{2,1} + r_{1,2} \times r_{2,2} + \cdots + r_{1,n} \times r_{2,n}$

   Note that the result of dot product is a scalar, not a vector.

6. Length: $|\mathbf{r}| = \sqrt{\mathbf{r} \bullet \mathbf{r}} = \sqrt{r_1^2 + r_2^2 + \cdots + r_n^2}$

   Just like a normal vector, the length of a RV denotes the distance between the point represented by the RV and the origin in the resource space. Similar to the dot product, the length of a RV is also a scalar.

7. Direction difference

   Being a vector, each RV has a definite direction in the resource space. We use $\mathbf{r}_1 \wedge \mathbf{r}_2$ to represent the angle between the two RVs in the resource space. Then the direction difference is defined as:

   $$D_{\mathbf{r}_1 \wedge \mathbf{r}_2} = 1 - \cos(\mathbf{r}_1 \wedge \mathbf{r}_2) = 1 - \frac{\mathbf{r}_1 \bullet \mathbf{r}_2}{|\mathbf{r}_1| \, \|\mathbf{r}_2|}$$

   Since each coordinate of $\mathbf{r}_1$ and $\mathbf{r}_2$ is greater than or equal to zero, the value of the direction difference D is between 0 and 1. Moreover, the greater the value of D, the greater the angle between $\mathbf{r}_1$ and $\mathbf{r}_2$ is. Note that to use this operation, the coordinates, representing different types of resources in practice, should first be normalized.

   In particular, when one of the two RVs is the total system resource capacity **RC**, say $\mathbf{r}_2 = $ **RC**, then $D_{\mathbf{r}_1 \wedge \mathbf{RC}}$ is the direction difference between $\mathbf{r}_1$ and the total system resource. We call this difference the resource deviation of $\mathbf{r}_1$, namely

   $$Dev_{\mathbf{r}_1} = 1 - \cos(\mathbf{r}_1 \wedge \mathbf{RC}) = 1 - \frac{\mathbf{r}_1 \bullet \mathbf{RC}}{|\mathbf{r}_1| \, \|\mathbf{RC}|}$$

$$= 1 - \frac{r_1 RC_1 + r_2 RC_2 + \cdots + r_n RC_n}{\sqrt{RC_1^2 + RC_2^2 + \cdots + RC_n^2} \sqrt{r_1^2 + r_2^2 + \cdots + r_n^2}}$$

When r_1 is the sum of resources used by all the applications in the system, then Dev_{r_1} represents the deviation of the total resource usage. Hence, to some extent, it can describe the balance of the consumption of different types of resources in the node system. The smaller the Dev_{r_1}, the better the system resource usage is.

Note that here **RC** is the resource capacity in a node system. It means the total amount of resources offered to user applications. Practically, it can be calculated by the total amount of system resources configured at the system initiation minus the various system resource cost for keeping the normal running of the system, or be measured at the idle state of the system, i.e., when no user applications running in the system. In general, the node system should configure **RC** in such a way that the ratio between the coordinates is proportion to the demand of applications on different types of resources, such that each type of resources can be well used. However, due to the diversity and complexity of applications, this ration cannot always be kept.

In addition, the following combinations of the operations are also applicable to the RVs:

- $\frac{r_1}{r_3} + \frac{r_2}{r_3} = \frac{r_1 + r_2}{r_3}$

- $(r_1 + r_2) \bullet r_3 = r_1 \bullet r_3 + r_2 \bullet r_3$

In conclusion, these RV operators are used in the adaptive admission control algorithm discussed in the following chapters and can simplify the procedures and the description of the algorithm.

3.3.2.3 RV and Resource Abstractions in NodeOS

As introduced in chapter 2, the AN research group has suggested five primary abstractions, i.e., thread pools, memory pools, channels, files and domains, to describe the system resources in [AN01]. The concept of RV does not conflict with these abstractions. The abstractions defined in [AN01] are used to encapsulate different system resources during the implementation of the NodeOS API, whereas RVs are used to describe the resource requirement, resource allocation and organization in an AN node from the perspective of active applications. RV is similar to all the resources allocated to a domain, however, this domain must be the highest level domain regarding an active application. In general, RV is a higher level abstraction than that defined in the NodeOS interface. To some degree, it can also be regarded that RVs define a resource abstraction between applications and EE, while the thread pools, channels etc. define a resource abstraction between NodeOS and EE.

3.4   Resource Usage in AN Node using RV

Based on RVs, the total system resources, resources required, reserved and consumed by applications as well as other concepts related to the resources in an AN node can be defined in the resource space. Generally, the following two points can be derived based on the RV concept.

First, due to the multi-dimensionality and complementarity, diverse resource combinations, each of which can be described using a RV, can satisfy the performance requirement of applications. In other words, the resource requirement of applications is a set of RVs. Furthermore, these RVs form a subspace in the resource space.

Second, the RV concept can reflect the usage of multiple types of resources in the node system directly and suggests also a method for allocating the system resources to applications, making full use of the system resources and keeping the different types of resources in balance. Figure 3.3 illustrates the relationship between the total resources in a node system and the resources used by each EE instance from the point of view of RV. From this figure we can see that only when the sum of the resources used by all the EE instances (Σr) coincides with the total resources in the system (**RC**), the system is saturated, i.e., no more applications can be accepted any more. In this case, the system resources have been fully used. In addition, if resources could be allocated to the applications along the direction of **RC**, namely the resource deviation of Σr is equal to zero, the system resources would remain balanced. In other words, the possibility that a new application cannot be served by the node only because of the shortage of one or several types of resources generally becomes smaller, and therefore the system resources can be better utilized.

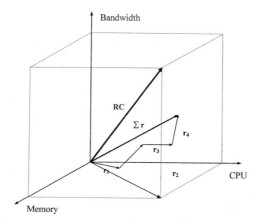

Figure 3.3: System resources vs. resource used by applications

The advantages of using RVs to describe the resources in ANs are summarized as follows:

- RVs reflect the resource usage characteristics in AN nodes, and can therefore be well used to solve the resource management problems in AN nodes. E.g., the coordinates of a RV denotes the number and amount of resources needed to fulfill the tasks of an application in an AN node. Moreover, a set of RVs, each representing a resource combination that can satisfy an application's performance requirement, can also be used to describe the complementarity. Of course, the resources that an EE instance maintains on behalf of an active application can be well described using RVs.

- RVs simplify the resource description. The multiple types of resources needed by active applications can be described using RVs, just like only one type of resource is needed in the classical network, this simplifies greatly the description of the resource management algorithm, e.g., the resource allocation or adaptation algorithm.

- RVs provide a basis for new resource allocation algorithms in AN nodes. Based on RVs, the concepts such as the available resource adaptation space, the resource usage deviation etc. can be established. These concepts are the basis of our adaptive resource management scheme in AN nodes, which tries to utilize the system resources optimally to satisfy the needs of as many applications as possible.

3.5 Summary

This chapter has analyzed the resource usage in AN nodes. On the one side, there is no fixed proportional relationship among the different types of resources consumed by an application. Therefore, various resources have to be discussed separately. On the other side, from the perspective of applications, there still exist some relationship among the different types of resources. Particularly, due to the mechanisms of resource sharing in an AN node, the resources allocated to one EE instance should also be considered as a whole entity. All these suggested that a new method for describing the resource usage in ANs is necessary.

Hence, we have introduced the resource vector, which involves both entire and individual aspects, as a method to describe the resource usage in an AN node. Each coordinate of a RV can reflect a type of resource needed by an application separately on the one hand. On the other hand, a RV can also be considered as an entirety to denote all the resources needed by the application. Furthermore, RVs can also simplify the description of resources. Some operations related to RVs have been described in this chapter.

The RV concept suggests also a method for using multiple types of system resources optimally. In case one or several types of resources is scarce, the coordinate values of the RV denoting the resource requirement of an application can be adjusted properly by using some algorithms and auxiliary operations of RVs, and simultaneously taking the attributes of applications into account. Thus, the system resource scarcity can be remedied to some degree. At the same time, the application can be served, and its performance requirement can remain in the permitted extent. Moreover, a proper adjustment algorithm can also impel the different types of available resources in the system to be kept in balance with its total capacity, i.e., along the direction of **RC** in the resource space. This is the basic idea of the work in this dissertation. In the following chapters, we will present our method for describing the adaptability of active applications and introduce the adjustment algorithm and mechanism that can keep the system resource usage in balance.

Chapter 4

Adaptable Resource Vector Space

Chapter 3 has introduced the concept RV to describe the resource usage in an AN node, including the resources allocated to an EE instance on behalf of an active application. This chapter discusses how to express the resource requirement of applications considering the possible complementarity among different types of resources. It begins with the discussion from the perspective of the adaptability of applications.

First the adaptability of a networked application is analyzed, and then the necessity of a method for describing the application adaptability is discussed. Following this the concept of the adaptable resource vector space (ARVS) is introduced, which can be used to describe the adaptability of an application. Then the definition of the elements constructing the ARVS is introduced and discussed. And finally, some work related to the adaptability of applications and its description methods is discussed.

4.1 Active Applications and Adaptability

The adaptability of applications has been studied for several goals, such as to degrade the performance of applications gracefully, or to model the time-varying characteristics of applications. In this work, we discuss the adaptability of applications from the perspective of its possible resource consumption combinations, with the aim to realize adaptations among different types of resources.

4.1.1 Adaptability and Resource Consumption

Two phenomena can be observed with regard to applications in networks. One is that some applications may have multiple ways of execution, each of which may bring about the same result. E.g., a video file can be first highly compressed and then transferred to another node, where it will be decompressed before delivered to the user; or it can be transferred directly to another user. In both cases, the receiver can get the wished video file, and furthermore, under some circumstances, the receiver may not distinguish clearly the difference of the videos obtained through the two methods. The other phenomenon is that users of some applications may sometimes accept the performance degradation of the applications. E.g., an image with smaller size than the original one may also be accepted by users sometimes.

The above two phenomena are viewed as adaptability of applications. In general, the adaptability means multiple ways of execution, each of which is decided by the environment conditions and can potentially be accepted by the users of applications. This involves three factors:

1. The application has multiple ways of execution.

2. All the results are acceptable. This may have two meanings. One is that the effectiveness of the multiple ways of execution may be the same. The other is that an execution may cause performance degradation, but this degradation is tolerable by the user of the application.

3. Which execution way is selected is decided by the conditions of the environment, particularly the resource status.

The interior reasons that cause the adaptability include:

- An application may have multiple algorithms, which have the similar effectiveness. Different algorithms may be selected under different conditions.

- An application may execute a series of algorithms. The sequence of the algorithms may produce different but acceptable effectiveness. Different stages may also have different algorithms.

- The algorithm of an application may have different settings, and therefore, brings out different acceptable results.

- The algorithm of an application may also consist of some modules for additional and more refined computation if the environment condition permitted.

The above possibilities may also be combined, thus at run time, form flexible ways of execution, resulting in the same or different acceptable performance.

From the perspective of resource consumption, the exterior reason that may cause the adaptability is the performance-resource mapping. Generally, an algorithm that produces an acceptable performance needs a certain amount of resources. However, different algorithms which may consume different amount of resources may also result in the same acceptable performance set. Therefore, basically it can be considered that the performance-resource mapping is not unique. From another point of view, it can also be considered that the amount of resources determines the selection of the algorithms.

In fact, the relationship between algorithms and the corresponding consumption of different kinds of resources has already been considered in some network systems and computation models. For example, both sparse matrices and hashtables can be used to represent certain data structures, however they need different memory to store the same amount of data and the CPU time for visiting the stored data is also different. In other words, a tradeoff exists between storage space and element access time. In addition, whether caching is used affects also memory occupation and network bandwidth and/or CPU time used for the information retrieval.

Hence, it is possible that by controlling the combinations of the amount and type of resources allocated to an application externally, different algorithms may be selected by the application internally, and at the same time, the performance of the applications remains in a tolerable scope.

As a summary, the adaptability of applications embodies the ability to trade off resource requirements over several dimensions, including resource type, amount, and output quality etc. The applications are able to compensate for a lower allocation of resources either by lowering the output quality, or by raising the demand for resources of another type. Such adaptability provides also flexibility to the resource management of the underlying system, which can select an operating point for an application that improves the whole system resource utilization while still ensures that the application can meet the predictable requirements of the users.

4.1.2 Adaptability of Active Applications

Active applications may carry their own programs together with the control parameters and other data which has to be processed or transferred in the packets to the networks. In principle, this increases the possibilities for adaptability. Furthermore, information about the adaptability characteristics may also be given to the network. In addition, active applications may also select different execution environments to execute their programs. This increases also the flexibility for applications to select different algorithms and perform adaptations. Hence, the adaptation capability of active applications could be much higher than that of the normal applications in the traditional IP-based networks.

In fact, active networking itself is somewhat a tradeoff between bandwidth and other computing resources, such as memory and CPU cycles. In [Wet99b], it has been argued that bandwidth resource is presumably the scarce resource at the node, because part of the rationale behind programmable networks is to trade off bandwidth for other computing resources, such as memory and CPU cycles.

In the following sections, we study the adaptability of active applications from the point of view of resource management in AN nodes. We focus on how to express the adaptability information so that this information can be transferred to the AN nodes.

4.2   Description of Adaptability

4.2.1   System-Side Adaptation

Support for making adaptation decisions can be provided by modules embedded in the application, or in the network node operating systems. The former can be viewed as the application-side adaptation. In this case, an application itself has to maintain the complex mechanisms for performance control and making adaptation decisions. In addition, information about the conditions of the network, or the current performance of the application have to be fed back to the applications. The applications can only decide how to execute the application continuously, namely whether and what kind of adaptations should take place according to the feedback information. Therefore, this method is time-taking.

The latter is also called the system-side adaptation. This method overcomes the disadvantages of the application-side adaptation. Because the network nodes maintain all the information needed for adaptations, including the various resource states and the decrease of the application performance. The time for transmitting this information to applications is also saved. Moreover, the system-side adaptation also lightens the burden of the applications. An application does not need to embed the mechanisms for monitoring the degradation of the performance and starting the corresponding adaptations, which increase greatly the complexity of the applications. Hence, we studied the use of the system-side adaptation methods in our work. However, using this method, network nodes must have detailed information about the adaptation capabilities of applications. Previous research has only considered models for describing the adaptation capabilities of applications in specific application domains, such as multimedia [HW96] [NS95], or applications demanding a fixed set of resources and service types [AAS97] [RJM+98]. These can, however, not satisfy our need for the adaptation for active applications.

4.2.2   Considerations about Adaptability Description

The method for describing adaptability characteristics must concern the following two issues. First, it must consider the internal features of the applications. Second, it must reflect the relationship between the applications and resources.

Considering the first issue, although the algorithms and their control parameters cause the multiple ways of execution of an application, which are the essence of the adaptability, it is however difficult to use these algorithms directly to describe the adaptability of an application. To list all the possible combinations of the algorithms and their control parameters for all applications that will be served by the networks is not a general and scalable method. As an alternative, an abstraction reflecting a possible resource usage of an application can be used, since the amount of resources that an application acquires can influence the application to select a suitable algorithm or control parameters on the one hand, and on the other hand, this abstraction correlates also with the environment parameters in the networks.

With regard to the second issue, the influence of the system resource variance on applications should be taken into account. Namely an abstraction related to both the applications and the variation of the system resource status should be used. E.g., an application may ask for more resources when it believes that currently there are much available resources in the system, which can be used to improve its performance.

Based on these considerations, we introduce the efficacy function and the cost function to establish the relationship between applications and system resources. According to these functions and the concept of RV, a closed subspace in the RV space can be constructed. Thus, when the resources that an application can get are inside this subspace, the performance requirement of the application can be satisfied. We call this subspace the adaptable resource vector space (ARVS). It stipulates the adaptable extent of an application, and can therefore, be used to describe the application adaptability. In the following sections, we first present the efficacy function and the cost function which are the basic elements of ARVS, and then discuss how to construct the ARVS to depict the adaptability of applications.

4.3   Efficacy Function and Efficacy Loss

The efficacy function and the efficacy loss are introduced to describe the performance and the performance degradation of an application in relation to different types of resources.

4.3.1   Efficacy Function

4.3.1.1   Definition

The efficacy of an application is defined as a metric for users to measure the general satisfaction degree with the application or the degree that the capability of the application can be presented, when the application executes in a certain environment, particularly under a certain resource condition.

The efficacy can be understood from both the intrinsic and the extrinsic side of an application. Intrinsically, an application has some power or capability to produce some levels of desired effects for its users through maintaining some algorithms. However, the power or capability of an application cannot always be fully expressed due to the limitation of the environment conditions, such as the amount of resources that can be allocated to the application. I.e., sometimes an algorithm that can provide better performance in general cannot execute under certain resource conditions. In other words, the environment resource conditions, more concretely, the resources that an application can acquire, determine the selection of the algorithms or the control parameters of the algorithms during the execution. Therefore, they affect the level that the capability of the application can be expressed.

Extrinsically, the efficacy can be viewed as a quality metric representing a satisfaction index of the application users, namely the degree of the application performance. This includes two meanings. One is that the efficacy can be represented as a separate quality of service (QoS) metric, denoting a general satisfaction degree to the application performance by the users, just like other QoS metrics, such as delay or jitter. In this case, the efficacy can be observed or tested by the users subjectively or objectively: for example, [ITU95] introduces a 5-level mean-opinion-score (MOS) method for measuring a video quality subjectively. The general quality of a picture can be measured by SNR (signal to noise ratio), PSNR (peak signal to noise ratio) and WSNR (weighted signal to noise ratio). The other meaning of the efficacy is that it can also be represented by a combination of other QoS metrics of the application, in order to express the general satisfaction degree with the application, instead of only a single aspect of the quality of the application.

In general, the efficacy of an application reflects the features of the application itself and depends on the resources that the application can obtain during its execution. Thus, when the application is determined, the efficacy varies only with the resources. I.e., the efficacy of an application can be considered as a function of different kinds of resources, namely as efficacy function, denoted as $Ef(\mathbf{r})$, where $\mathbf{r}$ is a resource vector. Obviously it is a multidimensional function of resources.

4.3.1.2 Features of Efficacy Function

Because of the complexity of applications and their algorithms as well as the control parameters for each algorithm, it is very difficult to give a general form of efficacy function to all applications. However, according to the above assumptions, some common properties of the efficacy functions associated with each resource dimension can be summarized as follows:

- Non-decreasing: typically, the increase of a type of resource does not decrease the application performance. This makes the efficacy function a non-decreasing function of each resource type.

- Piecewise-continuous: the efficacy function of an application may be not continuous in the range $[r_{min}, r_{max}]$, however, it can be piecewise continuous, such as shown in Figure 4.1 (b) and (c).

- With maximum extreme value: basically, the efficacy of an application will rise with the increase of the amount of resources consumed by the application. However, when the amount reaches a definite value, the performance will not increase.

- Always have definition in the bounded interval $[r_{min}, r_{max}]$.

Figure 4.1 illustrates some typical shapes of the efficacy function along each resource dimension, namely $Ef_j(r_j)$. For example, considering the resource network bandwidth, a linear efficacy function represents the case of an equal bandwidth adjustment. It serves data applications whose throughput increase equally with the increase of the network bandwidth within some extent and that are insensitive to the bandwidth variation over any particular range of bandwidth allocation. As another example, concave efficacy functions (e.g., Figure 4.1 (d)) model strongly adaptive applications that are sensitive to the bandwidth variations, especially if the bandwidth supply is close to the lower limit but not as much sensitive to the bandwidth changes at the end of the maximum bandwidth requirement. TCP represents an example of such an application. In contrast, the S-shape efficacy functions, as shown in Figure 4.1 (e), represent weakly adaptive applications that are sensitive to the resource changes when the bandwidth consumption is close to the maximum value, such as some video applications. The discrete efficacy function represents discretely adaptive applications, such as multi-layered MPEG video flows. Similarly, the adaptability of applications can also be described using the efficacy function along the CPU and memory resource axis respectively.

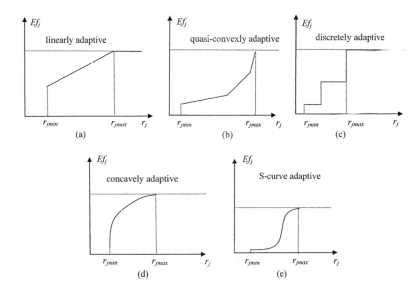

Figure 4.1: Typical efficacy functions along one resource dimension

From this figure we can see that in order to achieve a certain efficacy, an application must get a certain amount of each type of resources. We call this amount of resources the minimum sustained resources, i.e., r_{min}. In addition, the efficacy function has also a maximum value. At a definite point, no matter how many resources the application can get, the efficacy will not increase. We call these resources the maximum saturated resources, denoted as r_{max}.

The efficacy function maps the resources that an application can get in a network node system into the satisfaction degree of the user with the execution result of the application. It conveys the adaptability of an application, i.e., the resource consumption choices, to the networks, when the desired amount of resources cannot be provided.

4.3.1.3 Determination of Efficacy Functions

Generally, to acquire the efficacy function of an application is not easy. In our framework, we suppose that efficacy functions are defined by application providers. Users may specify the resource adaptability of an application by choosing an application-specific form of efficacy function and modifying customizable parameters of the efficacy function.

In the following we discuss two potential methods for the determination of application-specific efficacy functions. Both of them are based on the definition of the efficacy function.

Measurement Method

For some applications, efficacy functions can be represented by a quality metric denoting the general satisfaction degree by the application users. Hence, for these applications, efficacy functions can be acquired through the measurement method. The general steps are:

1. Select or define a method for assessing the quality of an application. Examples are to use the MOS method to assess a video application subjectively, or to use the SNR method to determine the level of satisfaction objectively. A combination of some parameters, such as the resolution and the size for an image presentation, can also be used to represent the satisfaction degree of users.

2. Select some sampling points, i.e., select some resource combinations represented by RVs, and measure the efficacy of the application under these sampling points. That means to assess the application quality using the selected method under several sets of resource combinations. After this step, several discrete points of the efficacy function can be acquired.

3. Get the efficacy values under other resource conditions using an interpolation method. Thus, the entire efficacy in a specified resource extent can be expressed.

As a result, an efficacy function can be got in the form of a polynomial. Obviously, this is only an approximation method for determining the efficacy function of an application. The precision of the efficacy function depends on the number of the sampling points.

As an alternative simplified method, the third step can also be omitted. In this case, the efficacy function can be formulated using the efficacy value at the sampling points of resources. We note that directly using staircase efficacy function leads to a complex combinatorial optimization problem according to our adaptive resource allocation algorithm. Therefore, we express the efficacy function as a quasi-continuous function instead of a discrete staircase shape. Consequently, it can be approximated using the piecewise linear curve, as depicted in Figure 4.2. Suppose K is the number of the sampling points, and the efficacy value at the sampling point k is e_k, then the efficacy function has the form:

$$ef_k(\mathbf{r}) = e_k + \frac{e_{k+1} - e_k}{\mathbf{r}_{k+1} - \mathbf{r}_k}(\mathbf{r} - \mathbf{r}_k), \qquad \mathbf{r} \in [\mathbf{r}_k, \mathbf{r}_{k+1}], \text{ and } k \in [1, K-1]$$

Note that using this method, in order to be able to model the efficacy function completely, $\mathbf{r}_1$ and $\mathbf{r}_K$ should select the minimum sustained resources and the maximum saturated resources. Otherwise the acquired result can only characterize the efficacy function partly. In addition, the choice of the number of the sampling points is engineering and application decision. The higher the number of the sampling points, the more accurate but complex the curve of the efficacy function is.

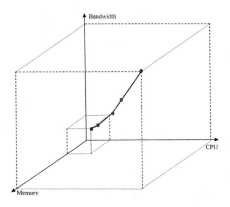

Figure 4.2: Linear approximation of the efficacy function

QoS Parameter Method

Another method to model the efficacy function of an application is to use other QoS parameters of the application. As mentioned above, the efficacy function represents a general satisfaction degree of a user with an application under different resource conditions. For some applications, a suitable way to assess the satisfaction degree of a user objectively is to use some other widely used QoS parameters of the application. The users can specify a combination of the QoS parameters according to their interest, to represent the efficacy function of an application.

Suppose application i has a x-dimensional QoS requirement. The relative weight of these QoS dimensions to the entire performance perception of the application constructs a vector $\mathbf{q_i} = [q_{i,1}, q_{i,2}, \ldots, q_{i,x}]$. $\mathbf{w_{i,x \times n}} = (w_{i,ab})_{x \times n}$ is a factor matrix, where n is the number of resource types related to the application performance. Each element $w_{i,ab}$ represents the relative dependence degree of QoS dimension q_a on resource r_b. Hence the efficacy function of application i could be denoted by $E_i(\mathbf{r}) = \mathbf{q_i} \bullet \mathbf{w_{i,x \times n}} \bullet \mathbf{r_i^T}$.

Through this method, an efficacy function is indeed expressed as a linear function of different kinds of resources with parameters depending on the QoS requirement of the corresponding application. In this sense, the efficacy function reflects the level of QoS degradation with the variation of resources acquired by the application.

In conclusion, an efficacy function can be acquired both through the measurement and by using the QoS parameter methods. In the former case, the efficacy is typically defined as a separate metric denoting the satisfaction level experienced by the users of an application subjectively or objectively. And normally a polynomial efficacy function can be acquired. This method can also be simplified, which results in a piecewise linear curve. In the latter case, the efficacy function is represented by a combination of some QoS parameters of an application, denoting the general satisfaction degree, and as a result, a linear efficacy function can be acquired.

4.3.1.4   Example of an Efficacy Function

In this section, we give an example of the efficacy function for an image application obtained by the analytical method.

Assume that an application has a two-dimensional QoS requirement with an image size of 800×600 pixels, and a compression level for JPEG of 90. The relative importance of both QoS parameters is $q = [0.8\ 0.2]$. The relative dependence for the QoS dimension image size on the network bandwidth resource b, CPU c and memory m are 0.7, 0.05 and 0.25 respectively. For the QoS dimension compression level, they are 0.1, 0.7 and 0.2 respectively. Thus, the factor matrix is

$$w = \begin{bmatrix} 0.7 & 0.05 & 0.25 \\ 0.1 & 0.7 & 0.2 \end{bmatrix}$$

and therefore the efficacy function becomes

$$Ef = q \bullet w \bullet r^T = [0.8\ 0.2] \bullet \begin{bmatrix} 0.7 & 0.05 & 0.25 \\ 0.1 & 0.7 & 0.2 \end{bmatrix} \bullet r^T$$

$$= [0.58\ 0.18\ 0.24] \bullet r^T, \qquad r \in [r_{min}, r_{max}]$$

where r_{min} and r_{max} represent the resource amount when the worst tolerable image size and compression level, and the best image size and compress level that the system can reach is provided respectively.

Suppose when the above QoS requirement can be satisfied, the needed resources are bandwidth 20000 bytes/s, CPU 300000 cycles/s, and memory 40000 bytes, denoted as [20000, 300000, 40000]. We use this resource demand to normalize the above linear function, and desire the efficacy function of the image application, namely

$$Ef = \frac{0.58}{20000}b + \frac{0.18}{300000}c + \frac{0.24}{40000}m \qquad (b, c, m) \in [r_{min}, r_{max}]$$

4.3.2 Efficacy Loss

A change of the amount of the resources obtained by an application will cause a variation in its efficacy. We call this efficacy variance the efficacy loss. Note that the efficacy loss can be negative, which means indeed an increase in the efficacy. We use the term efficacy loss, since in practice, normally each application has a desired resource combination, and the corresponding efficacy is the desired efficacy. When the amount of resources got by the application i deviates from the wished value $\mathbf{r}$ to the actual value $\mathbf{r}'$, we assume that there is some loss for the efficacy. In other words, we have an efficacy loss function $El_i(\mathbf{r}) = Ef_i(\mathbf{r}) - Ef_i(\mathbf{r}')$. For EE^j with m^j instances, the total loss is then $El^j = \sum_{i=1}^{m^j}(Ef_i(\mathbf{r}^j) - Ef_i(\mathbf{r}'^j))$. Correspondingly, the total loss in the active node due to the resource variation is $El = \sum_{j=1}^{l} El^j$.

4.3.3 Discussion

4.3.3.1 Efficacy vs. Quality of Service

The QoS is a generic term which takes into account several techniques and strategies in order to provide for applications and users a predictable service from the network and other components involved, such as operating systems. Generally it covers two aspects. One is a broad collection of networking techniques, such as scheduling, admission control, shaping, control on routing latency, control on performance and resource planning etc. These techniques involve the prioritization of the network traffic, in order to provide the various QoS for applications. The other aspect is the definition of the QoS using parameters. Since QoS covers different network elements, including applications, hosts or routers, and layers, such as application layer, transport layer or network layer, the definitions of the QoS parameters are diverse. For instance, normally the transfer delay, delay variance and packet loss are used to define the QoS in the network layer, and for the application layer, it is necessary to define the QoS according to the human perception and appraisal, e.g., using parameters of the refresh rate of a video stream (frame/sec), frame size and color resolution. Each of these parameters is normally called QoS dimension or QoS metrics.

The efficacy concept is related to the application layer QoS. First, the efficacy can be considered as a metric of the QoS, representing the general user satisfaction degree concerning an application. Therefore, it can be defined both separately just like the frame rate or color resolution, or as a combination of other widely used QoS parameters. Second, the efficacy is always cited as the efficacy function, which is a function of various types of resources, and denotes the satisfaction degree of a user under various resource conditions; whereas other QoS parameters are always regarded as the quantified measurement metrics of the user satisfaction. Usually they are not explicitly presented as a function of resources.

4.3.3.2 Efficacy Function vs. Utility Function

The concept of "utility" comes from economics, where it means the capacity of a commodity or a service to satisfy some human want. Utility functions have been widely discussed in the signal processing community in the form of rate distortion functions [Shan74] [OR98] for lossy video coding. In the networking community, utility functions have been discussed on an abstract level [Shen95] or as a model to formulate the network effect of TCP congestion control algorithm [KMT98] [LL99]. Furthermore, work such as [Khan98] has used the session utility function to model a multisession multimedia system, [BCL98] has used the bandwidth utility function to model the relative bandwidth preference of applications. The meaning of the utility functions in these projects are different more or less. More commonly, the utility functions define an objective of an adaptable system. In this sense, utility function and our efficacy function are the same. Both of them are a measure that quantifies the preference of a decision maker for one action over another. However, the utility functions address more related to the resource usage policies. Generally, they are derived based on issues such as revenue, fairness and priority. On the contrary, the efficacy function emphasizes particularly on the performance of an application. It acts more as a quality metrics of an application.

4.4 Resource Price and Cost Function

Besides the efficacy function, which represents the internal features of an application, an abstraction that connects an application with the environmental resource status should also be defined, in order to describe the adaptability of an application. Usually, the amount of resources that an application can obtain depends not only on its demand, but also on the system resource status, which also depends on the resource usage of other applications in the system. Therefore, this abstraction should also reflect the influence of one application on the system and thereby on other applications. For this reason, we introduce the resource price and the cost function as further elements.

4.4.1 Resource Price

In the economic world, the economic entities, such as commodities, customers, benefits and competition, are connected through selling and buying. Moreover, the amount of commodities that a customer can buy is limited by its own economic capability. This leads us to use the economic methods to describe the adaptation capability of an application. I.e., it can be regarded that applications buy resources from the network node systems. The capability for buying resources depends on the interior features of an application. Therefore, through competing for resource consumption, the relationship between applications and between applications and systems can be established.

The pricing system is the central part of an economic system. Basically two methods are used to establish a price system: fixed pricing and dynamic pricing. Fixed pricing means the price of commodities is fixed, it is normally determined according to the value of the commodities. Dynamic pricing means the price of commodities varies with the demand and supply of the commodities and reflects not only the value of a commodity but also the relationship among the economic entities. Hence, we use the dynamic pricing concept to represent the adaptation capability of applications. Once applications must pay for the resources used by them in a network node, they adapt their resource requirements according to the supply and demand of the resources in a network node, taking their needs and capabilities into account. The general idea behind the resource pricing is to provide incentives for applications to act in a manner compatible with other applications, and leading to a high total resource utilization in the network node system.

In a network node, the supply and demand of each kind of resources is not equal. Therefore, the price for each resource type is different. Similarly, the price for the same resource type but in different type of EEs can also be different due to the hierarchical resource organization in a node system. Based on the RV concept, we use the price vector $\mathbf{p}^j = \{p_1^j, p_2^j, \ldots, p_n^j\}$ to denote the price of the various resources in the EE^j.

The following issues should be taken into account during the establishment of the dynamic resource pricing system:

1. The resource price should reflect the resource status in a node system. As fewer the resources in the system are available, as higher becomes their price.

2. The resource price should be a convex increasing function over the amount of the already used resource r_{used}^j in the system. In other words, with a reduction of the available amount of a type of resource in the system, the corresponding resource price should rise sharply, in order to discourage the use of this type of resource. When a resource in the system is used completely, the corresponding price should go to infinite.

3. The total system resource capacity should also be considered in the resource prices.

According to these considerations, the resource price should be defined as a function of the resource load, i.e., the ratio between the already used amount and the total amount of resources, namely $\frac{r_{used}^j}{RC^j}$. Therefore, considering the first and second issues, the resource price in EE^j can be presented as:

$$\mathbf{p}^j = \frac{1}{\left(1 - \frac{r_{used}^j}{RC^j}\right)^x}.$$

Here the exponent x decides the rising speed of the price with the increase of the system resource load, the bigger the x, the sharper the price increases with the resource load. Figure 4.3 illustrates the price increase over the resource load under some selections of x.

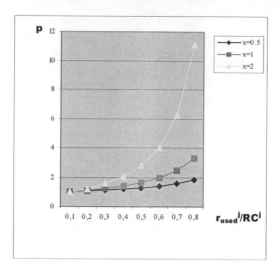

Figure 4.3: Definition of Resource Price

Balancing the increasing speed and the calculation complexity, we select $x = 1$. Therefore, the price becomes:

$$\mathbf{p}^j = \frac{1}{1 - \frac{r_{used}^j}{RC^j}}$$

where the coordinate $r_{t(used)}^j$ of $\mathbf{r}_{used}^j$ means the total number of resource type t used by all instances in EE^j, i.e., $r_{t(used)}^j = \sum_{i=0}^{m^j} r_{i,t}^j$.

4.4.2 Cost Function

Clearly the cost for an application that consumes resources is a function of the resource price and the amount of the resources it has consumed, i.e., $C = C(\mathbf{r}, \mathbf{p})$. Particularly, $\mathbf{p}$ is also a function of resources. Therefore, we call the cost also the cost function. Since both the resources and their prices are vectors, for the EE instance i in the type EE^j, the cost function can then be presented as $C_i^j = \mathbf{r}_i^j \bullet \mathbf{p}^j$. Note that the result is a scalar value.

4.4.3 Discussion

The system resource price changes when the resource load varies. Here, we have adopted the similar scheme like work in [KMT98] [LL99]. Concretely, when a new application arrives at the system, in case there are not enough resources for this new application, the resource prices in the system are calculated. According to our assumption, in fact applications know neither the system resource prices exactly, nor the changes of the resource prices. What an application knows is the highest cost that it can and has planed to afford. We think this is reasonable, because the goal of the application is to acquire the services from the networks. However, how many it can acquire from the network concretely depends not only on what the user wants to pay, but also on the time that he asks for the services from the networks, in other words, on the capability of other applications which are concurrent in the system.

In our scheme, for a specific application, the system resource price and the cost of the applications are only calculated when resource adaptations are performed. I.e., only when attempts to adapt the amount of different types of resources or reallocate the resources among different applications are made. And this occurs in order to serve a new application even though its requested resource cannot be reached. The price denotes the resource status rather than a monetary concept. It encourages or discourages an application to select a resource type in case of adaptations. From the perspective of the system, the cost indicates the efforts and concessions that the system makes for an application during the adaptation. And from the perspective of an application, the cost indicates the degree that the application can be adapted by the system: it prescribes the resources that the application can use. Apparently, resource cost does not have immediately anything to do with charging. However, since it implies the effort and concession that the system makes for an application during the resource adaptation, it should be taken into account for charging. This is, however, beyond the scope of our work.

4.5   Resource Limitation

As mentioned in section 4.3.1, an application has a minimum sustained resource r_{min}, in order to achieve a certain efficacy. r_{min} gives the minimum value of different kind of resources needed by an application. Similarly, it has also a maximum saturated resource r_{max}, which denotes the point where its efficacy will not increase further, no matter how many resources the application gets. In other words, consuming more resources are helpless, it increases only the cost.

r_{min} and r_{max} stipulate an resource extent that an application must acquire. When the resources obtained by an application are outside this extent, either the application cannot get any efficacy, or the application must pay more than necessary, which is not in the interest of the user of the application. Therefore, r_{min} and r_{max} construct in fact limitations to the adaptability of an application.

In order to address these adaptability limitations, we extract them from the efficacy function and use RV_L and RV_U to denote the r_{min} and r_{max} respectively in describing the adaptability of an application. RV_L and RV_U describe the lower and upper limit of various resources that the application must acquire, and they are the prerequisite for a useful execution of the application.

4.6   Adaptable Resource Vector Space (ARVS)

4.6.1   Definition

We have introduced the efficacy function, the cost function and the resource limitation separately. In practice, the efficacy function and the cost function intersect each other in the resource space and construct a common subspace. This subspace is cut off by two cuboids formed by RV_L and RV_U, and form a closed space. When the resources obtained by an application are in this space, the execution result of the application can be accepted by its user. Therefore, this space brings together the possibilities of the resource amount that an application can accept from the network nodes, and therefore can be used to represent the adaptability of applications for the network node systems. We call this space the adaptable RV space (ARVS) of the application.

4.6.2   Remarks about ARVS

First, ARVS is a closed space. This is reasonable since users can always tolerate the degradation of the applications only to some degree. The characteristic of being a closed space is also important for the networks. Although ARVS is a feature of applications, it is however provided to and utilized by the network node systems under the system-side adaptation. A closed space is helpful for a node system to accelerate the adaptation procedure.

Second, we use a continuous resource space ARVS instead of using discrete points to describe the adaptability. This is due to the fact that both the efficacy function and the cost function are continuous functions. Theoretically, any resource change Δr will increase or decrease the performance of an application, although practically due to the interior configurations of an application, such as the combinations of algorithms or the control parameters of algorithms, the possible resource amount that an application consumes maybe some discrete points in the resource space. However, taking the resource provisioning in point of view, it is reasonable to use a continuous space to describe the adaptability of an application. In addition, a continuous space makes it also possible to use some algorithms for solving continuous problems, which simplifies greatly the adaptation procedure.

4.7   Summary

This chapter has analyzed the adaptability of active applications and introduced the method of using ARVS to describe the adaptability of applications. The starting point for the introduction of ARVS is to establish a relationship between applications and the resources directly, so that suitable information can be provided to the network nodes for the system-side adaptation in order to increase the total system resource utilization of

AN nodes. For this purpose, we have introduced several concepts: the efficacy function of an application denotes the general satisfaction degree of a user with the application performance under different resource conditions. It reflects the application performance in relation to resources. The resource price denotes the resource status in network node systems. Together with the cost function, the preference to resources of an application is reflected in case of adaptations occur. The resource limitations reflect an adjustment extent for the various resources of an application.

ARVS describes the resource adaptation capability of an application. It contains sufficient and direct information about an application related to its resource usage. ARVS is used as the basis for the system-side adaptation during the admission control phase. We will discuss this problem in the next chapter.

4.8   Related Work

In this section, we discuss some work related to the study and the description of adaptation capabilities.

[Chang01] studies the adaptability of applications from the perspective of the configurations of an application. Chang believes that the execution flexibility of a tunable application is controlled by some control "knobs", which correspond to different configuration settings. Since these control knobs are typically implicit, Chang summarizes some configurations by exerting some parameters that can be manipulated from outside of applications to the application control structures. Thus, application adaptation becomes a switching from one configuration to another configuration by manipulating the explicit control parameters. The switching decisions are made and enforced by system components.

In that work, language annotations are used to describe the tenability of an application. These can be macro definitions in the source language or elements and attributes in the Markup languages, denoting the control parameters, components, transition functions and the QoS metrics of applications. At the runtime, a pre-processor analyzes the annotations and generates a tunable version of the application.

Chang exposes directly the alternative configurations of applications and covers the application specific knowledge of adaptability. However, the support of a pre-processor is needed to convert the annotations into the tunable versions. Moreover, during the execution, a monitoring agent plugged in by the pre-processor is used to monitor the resource status in the system. A resource scheduler obtains the resource information from the monitor, and computes the most appropriate configuration for the application. It controls the application execution by sending control messages to its steering agent generated by the pre-processor, which in turn sets up the application to execute the particular configuration. Furthermore, this approach has not given the concrete relationship between the applications performance or configurations and the various resources.

Both [Chang01] and our work address the description of the adaptability of applications. Both of them aim to adjust applications according to the resource conditions of the underlying system. Compared with the annotations in [Chang01], our ARVS presents a concrete relationship between the application performance and the consumed resources. Therefore, ARVS can be directly used in the adaptation decision and enforcement, no other translations are needed. In addition, during the description of the adaptability, we address also the relationship among different types of resources. On the contrary, although multiple types of resources are considered in [Chang01], they have not addressed the relationship among them.

[BCL98], [LC01] and other similar work of these authors use the bandwidth-utility function to model the adaptation behavior of an application. The bandwidth-utility function is defined to map the transmission bit rate of an application, referred to as available network bandwidth, into a "utility" value that represents the level of service quality satisfaction. Together with adaptation scripts, an application blueprint for adaptation is captured. The adaptation scripts are the principals for capturing the application-specific requirement on resource availability in terms of adaptation time scales and bandwidth granularities, e.g., what time scale and/or events should trigger an increase in bandwidth allocation and by how much.

In this sense, the bandwidth-utility has the same meaning as our efficacy function. Since the bandwidth-utility function abstracts only the bandwidth resource needs of an application, it can be considered as the dimension-wise efficacy function along the bandwidth axis. However, the main goal of the bandwidth-utility function is to model the "performance" of an application with the variation of the transmission bit rate of the application in a time scale, similar to the VBR (variable bit rate) service in ATM. Adaptations based on bandwidth-utility function stress when and how applications should react to the network bandwidth variations in the networks. The efficacy function is used to model the effect the "performance" when different amount of resources are allocated to an application. Here an implicit assumption is that active applications can finish their tasks in a time scale, within which an application itself does not have any resource demand fluctuation. Adaptations based on the efficacy function address whether an application will or can continue its execution if the amount of the resources acquired by it is different than those it requested. In other words, the ultimate goal of the efficacy function is to determine the possible amount of resources that an application can accept. Furthermore, since the efficacy function is a function of multiple types of resources, it is a surface in the resource space. Whereas the bandwidth-utility function is only a curve in the bandwidth-utility plane.

Moreover, besides the efficacy function, we use also the cost function and the resource limitations, namely ARVS, to describe the adaptability. Both the cost function and the resource limitations play an important role in the resource adaptation, just like the efficacy function. We will discuss the usage of ARVS in chapter 5. However, work in [BCL98], [LC01] etc. uses only the bandwidth-utility to capture the adaptive nature of mobile applications in terms of the range of bandwidth over which applications prefer to operate. Adaptation scripts only complement bandwidth-utility functions by capturing application-specific "adaptation time scales" and "bandwidth granularities". The utility-based network control realizes utility-based max-min fairness. Namely when the available bandwidth in the system fluctuates, the system adjusts bandwidth to applications such that each application has the same utility.

In summary, the bandwidth-utility concept reflects the effect of the bandwidth variation during the application execution on the benefit of the applications and the total system. Our ARVS concept generalizes the relationship between the application performance and the corresponding resource consumption. It addresses the contribution of each resource type to the application performance, and thereby reflects also the relationship among different resource types.

Chapter 5

Adaptive Resource Admission Control in Active Network Nodes

In the previous chapters, we have discussed the resource requirement and adaptability of active applications, and introduced how to describe them. In this chapter, we will discuss how these information are used by the AN nodes through introducing the adaptive resource admission control mechanism used in the AN nodes.

We begin with a brief introduction about the background information regarding the optimization problem which is the basis of the adaptation algorithm. Then we present the resource admission control mechanism, in which adaptations may occur. And finally, an evaluation of the adaptation algorithm is presented.

5.1 Optimization Problems

Optimization is an activity that aims at finding the best (i.e., optimal) solution to a problem. For optimization to be meaningful there must be an objective function to be optimized and it must have more than one feasible solution, i.e., a solution which does not violate the constraints to the problem. The term optimization is usually applied when a number of solutions permit the best one to be chosen by inspection using an appropriate criterion. In mathematical terms, the formulation of an optimization problem includes the following three basic ingredients:

- An objective function $Q = f(x_1, x_2, ..., x_n)$, which is to be minimized or maximized.

- A set of decision variables $x_1, x_2, ..., x_n$, which decides the value of the objective function.

- A set of constrains, usually in the form $G_i(x_1, x_2, ..., x_n)$ greater than or equal to a constant $O_i, i = 1, 2, ..., m$, which are limitations to the variables. For some problems, there may be no constraints to the variables.

In addition, criteria are also always used to decide if an optimal solution, namely a set of values of decision variables $x_1, x_2, ..., x_n$ that satisfies the constrains and for which the objective function attains a maximum or minimum, is found and/or the optimization procedure should be terminated.

In a word, the goal of optimization is to find a set of values for variables that minimize or maximize the objective function while satisfying the constraints.

5.1.1   Basic Principle

Since very few optimization problems can be solved analytically, in most practical cases appropriate computational techniques are used to solve the optimization problem. The basic principle for solving optimization problems computationally is probe and iteration. First a set of initial values for the variables is selected and used to probe the value of the objective function. And then the criteria are used to check if the current value of the objective function is optimal under this variable set. This process is called one iteration. Then a certain strategy is used to modify the initial values of the variables according to the probe results of the objective function in the former iteration, and the new values are used to continue to probe the objective function. The iteration continues until a set of values is found, which can pass through the check of the criteria.

Normally, the strategy used to modify the values of the variables at the kth iteration for finding the $k + 1^{th}$ iteration includes two steps, namely,

- Finding a search direction p_k, along which the optimal point may lie.

- Choosing a step length α_k.

Then the next probe value becomes $x_{k+1} = x_k + \alpha_k p_k$. The iteration continues until x_s is found, which makes the objective function minimum according to the criteria. Here s is called the iteration times.

In practice, there are many methods for solving the optimization problems, such as Gradient method, Newton method, and Lagrange method etc. [Fle86]. They use different techniques to choose the search direction, such as steepest descent, Lagrangian function etc., and the speed for reaching the optimal point and the complexity of the methods are different. Normally the convergence speed and the precision, i.e., the number of the iterations and the difference between the solution and the actual minimum or maximum value of the objective function, are two key issues used to evaluate an optimization method. Since besides the iteration number, the most usually used criteria for judging the termination of an optimization procedure are the difference between two continuous iterations. If the difference is small enough, the iteration will terminate. However, this value may still not be the optimum.

According to the features of the objective functions and the constraints, the optimization problems are divided into different subclasses. Figure 5.1 illustrates a kind of classification summarized in [NEOS]. Here, the optimization problems are divided into continuous and discrete problem according to the values which are permitted for the variables. For the continuous problem, according to whether the constraints to variables exist, it is divided into constrained and unconstrained problem. And the constrained problem is further divided into linear, nonlinear, bound constrained etc. problem according to the characteristics of the objective functions and constraints. Note that optimization is also called programming in many literatures[1], since the optimization problem is normally solved through programming techniques.

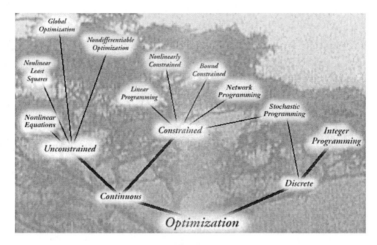

Figure 5.1: Optimization tree summarized in the NEOS server

Generally, there exist already some techniques specialized at solving definite optimization problems, such as Sequential Quadratic Programming, Augmented Lagrangian Method for the nonlinear constrained problem, Newton and Gradient-Projection method for the bound constrained optimization problem etc. Some software packages implementing these techniques for solving various optimization problems are also developed. The NEOS server for optimization has a good survey about the optimization techniques and software [NEOS].

In the following sections, we take a brief look at the continuous optimization problem, which is involved in our work. Note that we use the normal font to denote vectors or matrixes which have multiple components, whereas the cursive font to denote the components of vectors or matrixes.

---

[1]The term "programming" comes from 1940-1950, which means "optimization".

5.1.2   Linear and Nonlinear Optimization

When both the objective function and the constraints are linear functions, the optimization problem is called linear optimization or linear programming(LP). LP is usually stated as:

$$minimize \quad cx \tag{5.1.1}$$
$$subject\ to \quad Ax + a = 0 \tag{5.1.2}$$
$$Bx + b \leq 0 \tag{5.1.3}$$

where x is the vector of variables to be solved, A and B are matrixes with known coefficients, and c, a and b are coefficient vectors. The formula (5.1.2) and (5.1.3) are called equal and unequal constraints, respectively. Usually A or B has more columns than rows, and $Ax + a = 0$ is therefore quite likely to be under-determined, leaving great latitude in the choice of x with which to minimize the objective function cx. So far, there exist already very efficient solvers for the LP problem.

A Nonlinear Programming (NLP) is a problem that can be put into the form:

$$minimize \ f(x) \tag{5.1.4}$$
$$subject\ to \ g_i(x) = 0 \ \text{for } i = 1, ..., m_1, \ \text{where } m_1 \geq 0 \tag{5.1.5}$$
$$h_j(x) \leq 0 \ \text{for } j = m_1 + 1, ..., m, \ \text{where } m \geq m_1 \tag{5.1.6}$$

The formula (5.1.5) and (5.1.6) are the equal and unequal constraints, respectively. m_1 and $m-m_1$ are correspondingly the number of the equal and unequal constraints. The formula (5.1.5) and (5.1.6) may also not exist. In this case, the NLP becomes an unconstrained problem.

The main methods for solving the unconstrained NLP problem include the Gradient method, where the search direction $p_k = -g_k$; the Newton method, where $p_k = -H_k^{-1}g_k$; here g_k and H_k are the Gradient and Hessian of $f(x)$, respectively. Namely suppose x=$(x_1, x_2, ..., x_n)$, then

$$g_k = \nabla f(x) = \begin{bmatrix} \frac{\partial f}{\partial x_1} \\ \vdots \\ \frac{\partial f}{\partial x_n} \end{bmatrix} \qquad H_k = \nabla^2 f(x) = \begin{bmatrix} \frac{\partial^2 f}{\partial x_1 \partial x_1} & \cdots & \frac{\partial^2 f}{\partial x_1 \partial x_n} \\ \vdots & \cdots & \vdots \\ \frac{\partial^2 f}{\partial x_n \partial x_1} & \cdots & \frac{\partial^2 f}{\partial x_n \partial x_n} \end{bmatrix}$$

In practice, since the evaluation of the second derivatives for the Hessian is computationally expensive, the Hessian matrix is always approximated using other methods and the Newton method is therefore called the Quasi-Newton method [Fle86].

Since generally the constrained NLP is a difficult problem, researchers have identified some special cases and found efficient solvers specialized for them. A particularly well studied case is the one where all the constraints $g(x)$ and $h(x)$ are linear. The name for such problems, unsurprisingly, is "linearly constrained optimization". If the objective function is quadratic, the problem is called Quadratic Programming (QP).

The basic idea for solving the constrained NLP is to form a new objective function, which includes the constraints, i.e., the formula (5.1.5) and (5.1.6), and solve the new objective function using the unconstrained NLP method. For example, the Quadratic Penalty Method adds a quadratic penalty term to the objective function, i.e.,

$$minimize \quad Q_c(x) = f(x) + \frac{c}{2}\left[\sum_{i \in E} g_i(x)^2 + \sum_{i \in E} min(0, h_i(x))^2\right]$$

to obtain a solution $x^*(c)$, c is the penalty parameter. Then increase the penalty parameter and use the solution as the starting point for the next problem. In the limit there holds for the optimum of the constrained problem $x^*=\lim_{c \to \infty} x^*(c)$. Other methods for solving the constrained NLP includes, e.g., the Logarithmic Barrier method, the Exact Penalty method, the Newton method, the modified Simplex method and the Sequential Quadratic Programming (SQP) [Fle86].

If the variables of an optimization problem are limited in a bound, the problem is called bound constrained optimization, which can be stated as:

$$minimize \quad f(x)$$
$$subject\ to \quad l \leq x \leq u$$

Generally, the bound constraints can be considered as the unequal constraints; there are also some methods specialized on the bound constrained optimization problem, such as [BLN+94] and [NY97].

5.1.3   Multi-Objective Function Optimization

In practice, rarely does a single objective with several hard constraints represent adequately the problem being faced. More often are multiple objectives that must be tradeoff in some way, since the optimal solution for one objective is usually not the optimal solution for all other objectives. This kind of problems is called multi-objective optimization problem (MOP). A mathematical description of such problems is as follows:

$$minimize\ F(x) = \begin{bmatrix} f_1(x) \\ f_2(x) \\ \vdots \\ f_p(x) \end{bmatrix},\ where\ p \geq 2 \tag{5.1.7}$$

$$subject\ to\ g_i(x) = 0\ for\ i = 1, ..., m_1,\ where\ m_1 \geq 0 \tag{5.1.8}$$

$$h_j(x) = 0\ for\ j = m_1 + 1, ..., m,\ where\ m \geq m_1 \tag{5.1.9}$$

$$l \leq x \leq u \tag{5.1.10}$$

Note that, because F(x) consists of multiple functions, if any two components of F(x) are competing, there is no unique solution to this problem. Therefore, the concept of noninferiority [Zadeh63] (also called Pareto optimality [Cen77] [CP67]) is always used to characterize the objectives. A noninferior solution is one in which an improvement in one objective results in a degradation of another[2].

The basic strategy for solving MOP problems is to enable a natural problem formulation. Usually the evaluation function method is used. I.e., with the help of some geometric observation or background knowledge of the MOP problem, an evaluation function for the problem is constructed based on the original objectives. Thus, the multiobjective problem is converted to a single-objective problem. In other words, the multiple objectives are combined into one scalar objective whose solution is a Pareto optimal point for the original MOP.

Generally, methods for constructing the evaluation function include ideal point, weighted sum, minimax and goal programming.

Ideal point method

Using this method, first p single objective problems are solved, i.e., *minimize* $f_j(x)$, $j = 1, ..., p$. Suppose the solution is F* $=(f_1^*, f_2^*, ..., f_p^*)^T$, which is an ideal but almost impossible to reach point. Thus, under certain criteria, a point that is most approximate to F* can be searched and regarded as a solution of the MOP. A direct method to construct the evaluation function is:

$$minimize \quad \varphi(F(x)) = \sqrt{\sum_{i=1}^{p} (f_i(x) - f_i^*)^2}$$

and its solution x* is regarded as an optimal solution of the MOP.

Weighted sum method

Using this method, one can give more consideration to the relative important objectives. The evaluation function is constructed as follows:

$$\sum_{i=1}^{p} \alpha_i f_i(x), \ \alpha_i > 0, \ i = 1, 2, ..., p$$

The weight parameter α_i decides the relative importance of the objective functions. It is up to the user to choose appropriate weights according to the concrete optimization problem.

Minimax method

The essence of this method is to find a "best" solution under the "worst" situations.

---

[2]A point x*∈C is said to be Pareto optimal or a efficient solution or a non-dominated or a non-inferior point for MOP if and only if there is no x∈C such that $f_i(x) \le f_i(x^*)$ for all $i \in (1, 2, ..., n)$, with at least one strict inequality. Pareto optimal points are also known as efficient, non-dominated or non-inferior points.

Therefore, the evaluation function is constructed as follows:

$$minimize \ \varphi(F(x)) = min \ max_{l \leq j \leq p} \ f_j(x)$$

Goal programming

Another method for solving MOP is to minimize the primary objective $f_j(x)$, and expressing other objectives in the form of inequality constraints, i.e.,

$$minimize \quad f_j(x)$$
$$subject \ to \quad f_i(x) \leq \epsilon_i, \ i = 1, ..., p, \ i \neq j$$

This method is especially useful if a user can afford to solve just one optimization problem. However, it is not always easy to choose the appropriate "goals" for the constraints. Goal programming cannot be used to generate the Pareto set effectively, particularly if the number of objectives is larger than two.

5.2 Adaptive Resource Admission Control Mechanism

In order to satisfy the performance requirements of applications, network node systems should guarantee the resource availability requested by applications. Resource admission control checks if there are enough resources in the node system for a new application; therefore, it is a direct method for providing resource guarantees for applications. However, due to the multi-dimensional resource requirements of active applications, admission control at an AN node must check several types of resources simultaneously. Hence, situations may arise that an application is rejected because of the shortage of only one kind of resource although there is still a surplus of other types of resources. This is not beneficial to both the applications and the node system.

Therefore, we adopt an adaptive admission control mechanism in the AN node systems. Our goal is to admit as many applications with resource guarantees as possible and at the same time achieve a high utilization of all kinds of the resources in the system. The adaptive admission control mechanism can adjust the different kinds of resources requested by an application, or redistribute resources among the existing applications in the system during the admission control procedure, making use of the adaptability information of the applications.

5.2.1 General Principle

The adaptive admission control mechanism is based on the prerequisite that an active application informs the AN nodes of its desired resource requirement and adaptability together with its execution requirement in the AN nodes. In other words, it carries the desired resource requirement r_d and the adaptability information ARVS in the first packet to an AN node. The AN node system tries to satisfy r_d directly considering that the amount of the resources required by the applications can best satisfy their performance need. I.e., the node system performs admission control according to the r_d and the system available resources. If r_d cannot be satisfied directly, the node system adjusts the coordinates of r_d, and/or the amount of various resources allocated to other active applications in the system. The rules followed during the adjustment include:

- Remain inside the ARVSs, namely the resource vector r finally allocated to an application must be inside the given ARVS, otherwise the performance of the application cannot be guaranteed.

- Keep the system resources as balanced as possible. This means the total resources used in the system R_{used} should be kept along the direction of the system resource capacity RC as much as possible after the system resources are redistributed.

Figure 5.2 depicts the general principle of the adaptive admission control mechanism. Each application carries its resource request information in its first packet to the AN, including the desired resource requirement r_d and the adaptation capability ARVS. When the packet arrives at an AN node, the node checks if the required resources can be met by the system. If yes, the node accepts the application and sends a positive response in the first response packet to the application. If not, the node tries to adapt the amount of different kinds of resources that the application requests, or redistributes the resources among different applications according to the ARVS information provided by the applications. If the adaptation is successful, the node sends a positive response to the application, and notifies the application about the concrete amount of resources it can get at this moment. In both cases, namely with or without adaptation, the node stresses that it keeps the right that the assigned resources might be changed. Because the adaptations related to this application may become necessary due to the arrival of other applications. If the adaptations cannot be performed successfully, the node must reject the application. On receiving the positive or negative response from a node, the applications can decide themselves what to do further, e.g., stop the execution or probe later etc.

Note that in Figure 5.2, a general presentation of the ARVS is illustrated. In practice, the desired resource r_d, the worst efficacy value WE and the highest cost HC can also be discarded by the application. In this case, the AN nodes may calculate the desired resource requirement for an application according to the ratio of the efficacy and the cost, e.g., when the efficacy is 5 times larger than the cost, the needed resource is ideal. Similarly, as default, the AN node may also suppose that the maximal efficacy loss and cost increase for the application are both, e.g., 5%.

The characteristics of this mechanism are summarized as follows:

1. All the information needed for the admission control and adaptation are sent to the AN nodes in one block, no other control/signaling information exchange between the AN nodes and applications is needed during the execution of the applications.

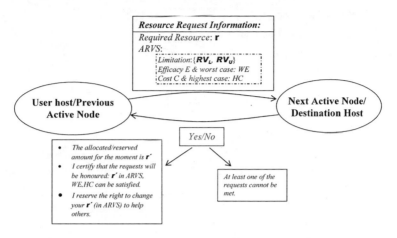

Figure 5.2: In-band r_d and ARVS transmission

2. The "in-band" information transfer mechanism is used. I.e., the resource request and the adaptability information are carried together with other data or program in the active packets, and the positive or negative confirmation is also integrated in the first response packet sent back to users. No separate control packets and messages are needed.

3. This mechanism adopts the immediate admission control and adaptation technique. Each AN node performs admission control on applications right after they arrive at the node. Adaptation is done immediately if needed.

This mechanism requires the active applications to pack their resource request and adaptability information in the first packets to an AN node. And the applications must also understand the result of the admission control integrated in the first response packets to them. This may increase some complexity for the development of applications. However, we believe that this mechanism is much more simple than introducing resource status control and adaptation mechanism in applications. In addition, this mechanism is based on the assumption that the application itself together with the underlying node system has a reliable scheme for transmitting the first packet containing the r_d and ARVS information, e.g., the application can be implemented based on TCP or adopt some retransmission or numbering scheme. This assumption conforms however to the goal of application developers and users.

In the following, we discuss the key points for implementing the above adaptive admission control mechanism in AN nodes, including how to present the resource request and adaptability information in the active networks, and how to perform the resource adaptations during the admission control.

5.2.2 Presentation of RV and ARVS within Networks

In the previous sections, RV and ARVS have been discussed. A more practical question related to these methods is how they can be represented and used within the adaptive admission control mechanism. According to chapter 3, a RV can be expressed using its coordinates. And depending on section 4.4, the cost and the cost increase can be calculated in the AN nodes according to the used resources and the available resources in the node system. Therefore, the main problem becomes how to express the efficacy function.

In the following, we first discuss the presentation of efficacy functions, then have a look at how to present the r_d and the ARVS from applications to network nodes or between two network nodes.

5.2.2.1 Efficacy Function

According to the assumptions in section 4.3, the efficacy function is a polynomial of different kinds of resources with some limitations. Since the efficacy functions vary with the applications and do not have a definite form, we adopt a general method to express the efficacy functions in the packets traversing through the networks. To simplify the explanation of our method, we use an arbitrary polynomial as an example of the efficacy function.

Suppose an efficacy function has the form of $5b^6 + 3c^5 + 2b^2c^3m + 4m^2 + 5$, b, c and m denote the network bandwidth, CPU and memory resource, respectively, then this function will be packed into a packet traversing the network as shown in Figure 5.3.

Since a polynomial consists of some terms involving the variables representing different resources, we express the polynomial using the terms, each of which has the same data structure. As shown in Figure 5.3, each term is depicted using a seven-byte array, 4 bytes for the coefficient of the term, standing for a float[3]; and the other three bytes for denoting the exponents of the variables[4], each byte represents an exponent for one variable. The total length of the efficacy function depends on the number of the terms of the polynomial. Although there may be some wastes for the constant term, in which the exponents of all the three variables are zero, we think using the same term structure makes it easy for the AN nodes to reconstruct the efficacy function.

5.2.2.2 Resource-Vector (RV)

Since the structure of a RV is relative definite, in order to reduce the overhead for expressing RVs, we adopt the fixed length and format method to carry a RV within an active

---

[3]To reduce the overhead for transmitting the efficacy function cross the network, we use 4 bytes to denote a float instead of expressing it as a character string. The first three bytes are used for representing the integer part of the float, and the last byte denotes the decimal part with two digitals and the sign. Therefore, the coefficient is now between -16777216.99~16777216.99.

[4]We think the exponent of each variable between -127~128.

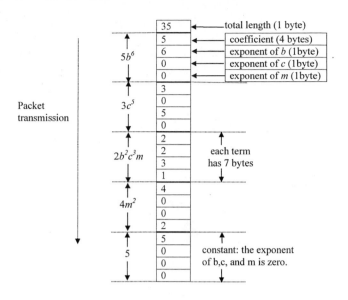

Figure 5.3: Transmission of efficacy function in the networks

packet. Figure 5.4 depicts the format of a RV.

Using this method, no indication for the resource type and length is needed, but the bandwidth, CPU and memory resource has a fixed order, and each of them is denoted using 4 bytes, and have an extent from 0 to 4294967295.

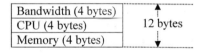

Figure 5.4: Transmission of efficacy function in the networks

5.2.2.3 ARVS

Based on the description of efficacy function and resource vector, an ARVS can be simply expressed. To keep the total length of ARVS as small as possible, we arrange the parameters of an ARVS in a fixed order, as shown in Figure 5.5. The highest cost and worst efficacy are represented in the percent of the desired value using two bytes respectively, varying from 0.01 to 99.99. Since the length of the efficacy function is variable, the whole length of the ARVS is also variable. As mentioned in section 2.2.4, the active network encapsulation protocol (ANEP) has defined the general format of the active packets, whose header consists of a set of fixed fields and some options. Therefore, we put the ARVS in the header of the active packets, as an option field of the ANEP, which has the format of Type/Length/Value. We will explained this further in section 6.1.2. A detail description about the format of the ARVS option can be found in appendix A.

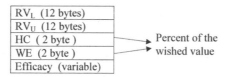

Figure 5.5: ARVS format

5.2.3   Admission Control Algorithm

Now we present how adaptations are performed during the admission control procedure in an AN node system.

When there is a new application request in an AN node, the node system checks the various resource states in the system to decide if the desired resource requirement r_d can be satisfied. If yes, the application is accepted immediately and r_d and ARVS carried in the packet is registered in the table CommonARVS in system. If not, the system adjusts the coordinates of r_d according to the corresponding ARVS and the available system resources. Or other applications might be adapted, i.e., the coordinates of the RVs related to other ARVSs already registered in the CommonARVS might be adjusted, with the aim to accept the new application under the current system resource states. In order to reduce the resource re-allocation and correlation among applications to make the system as simple as possible, the following steps are taken during the adaptation.

1. Adaptation inside the new application, namely to adapt the amount of the different types of resources that can be accepted by the application according to the current system resource states. If this step fails, the next step is performed.

2. Adaptation among the applications belonging to the same type of EE, that means to re-distribute resources among the applications belonging to the same type of EE. If this still fails, the third step is done.

3. Adaptation among the applications belonging to different types of EE, i.e., to re-distribute resources among different types of EE.

Figure 5.6 illustrates the general admission control and adaptation algorithm. Suppose that at one time point, there are k-1 applications in a node system, and the resources consumed in the system is $\mathbf{Rused_{k-1}}$, the available resources in the EE are then $\mathbf{Rav_{k-1}}$=RC-$\mathbf{Rused_{k-1}}$. At this time, the kth application arrives with the desired resource requirement $\mathbf{r_{kd}}$. If not $\mathbf{r_{kd}} \leq \mathbf{Rav_{k-1}}$, which means that at least one type of the resources in the system is not sufficient for this application, the adaptation within the $\mathbf{r_{kd}}$ begins first, namely to adjust the length of the coordinates of $\mathbf{r_{kd}}$ according to the coordinates of $\mathbf{Rav_{k-1}}$ and ARVS$_k$. That means a $\Delta\mathbf{r_k}$, some of whose coordinates are negative, should be found, so that $\mathbf{r_k}=\mathbf{r_{kd}}+\Delta\mathbf{r_k}$ satisfies (1) $\mathbf{r_k} \leq \mathbf{Rav_{k-1}}$ and (2) $\mathbf{r_k} \in$ARVS$_k$.

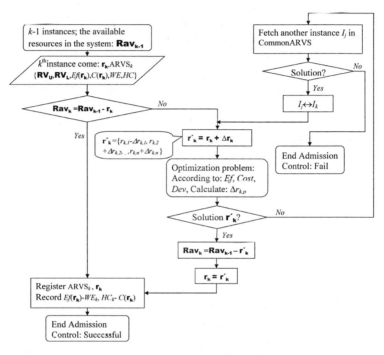

Figure 5.6: Admission control algorithm

We use the optimization method to find a solution for $\mathbf{r_k}$. We will discuss the objective functions, the constraints as well as the solving techniques in detail in the following section.

The above procedure is adaptation done within an application. In case no solution for $\mathbf{r_k}$ can be found in a definite time, which means the lacking resources cannot be compensated by other types of resources under the current system resource status, adaptations among different applications are performed. In order to speed up the optimization process, we reduce the number of variables as much as possible. Hence, we make the assumption that the resource requirement $\mathbf{r_{kd}}$ of the application k will be satisfied if another application in the same EE can be found, whose performance can still be met after the resource adaptation. That is to say, the missing amount of resources for the new application will be compensated by the same type and amount of resources from another application.

Apparently, the time spent on admission control correlates with the number of adaptations performed. In order to increase the possibility of a successful adaptation, during the second step, we use the following strategies to search a candidate application k' that has already been admitted in the system:

- $E(\mathbf{r}_{k'})$-$WE_{k'}$ is maximum, which means the efficacy of application k' is still far away from the worst value (WE) that the application can tolerate.

- $HC_{k'} - C(\mathbf{r}_{k'})$ is maximum, which signifies that the cost of the application is still much below the highest value (HC) permitted by the application.

- There is still enough unoccupied memory resource for the application. Because unlike the network bandwidth and CPU resource, which are counted using bytes per second and bytecodes per second respectively, memory is allocated e.g., to objects, threads and buffer of the application, and is counted using bytes. Therefore, the adaptation result for memory must at least be greater than the amount already used. Considering that adaptation needs some time, during which more memory may be allocated by the application, for the moment, we set the value of unused memory to 50%, i.e., $(m_{alloc} - m_{used})/m_{alloc} > 50\%$. However, this value needs to be studied further.

For the moment, we treat the efficacy and cost, i.e., the term $Ef(\mathbf{r}_{k'})$-$WE_{k'}$ and $HC_{k'}$-$C(\mathbf{r}_{k'})$ to be of same importance for adaptation. Therefore, the sum[5] S of these two terms is recorded in the system, when an application is accepted after the admission control. Hence, if the first step of adaptation, i.e., adaptation within the new application fails, we select the application with the largest value of S, and check if its memory usage is still within the permitted value for adaptation. If yes, the adaptation within this application is performed. If the adaptation can still not succeed, we select the application with the second largest S, check its memory usage and perform the adaptation and so on.

We have designed a table called CommonARVS for the selection of applications during the second step adaptation. It stores the ARVS information of each application in the system in descending order on S. After an application passes the admission control and is accepted by the system with the resource $\mathbf{r}_k$, the current value of the efficacy and cost difference, namely $Ef(\mathbf{r}_k)$-WE_k and HC_k-$C(\mathbf{r}_k)$, and the sum S of them are calculated. Then the ARVS information of this application is inserted in the table CommonARVS, using the binary search method to locate its position. The time spent on the position location depends on the number of the active applications running in the system. We will compare

---

[5]In practice, a weighted sum of the two terms can also be used. Users or the system can specify the weights.

this cost with an optimization procedure in section 5.3 to explain that adopting such a strategy may reduce the overall time for admission control.

Normally, the second step will end when all the applications registered in the table CommonARVS have been checked and selected for optimization probes. However, in order to control the total time spent on admission control, the number for optimization probes should be limited. Since the time cost for one optimization is system dependent, this number should be set by the node systems according to some statistics.

If a solution can still not be found after two steps of adaptation, the third step may take place. In other words, some resources can be borrowed from other EEs (say, in an "emergency case"). In principle, this procedure is the same as that for the second step, e.g., several CommonARVS tables can be used. However, in practice it may be complicated due to the need of information exchange among different EEs. So far, our AN node system supports only one type of EE; therefore, we have not implemented this step of adaptation.

After a candidate application is found in step 2 and 3, the same optimization method as used in step 1 is employed during the adaptation to find a suitable value within the ARVS of this application.

5.2.4 Optimization Procedure

Now we discuss how to use the optimization method to find a solution r_k inside the ARVS of the application k with the desired resource requirement r_{kd}. As stressed above, the application k can be the new arriving application or a helpful application found during step 2 and 3.

5.2.4.1 Objective Function and Constraints

In order to determine the objective functions and constraints, we first summarize the prerequisites for the optimization as follows:

- The system resource condition must be guaranteed. I.e., the new found value r_k must satisfy $r_k \leq Rav_{k-1}$

- The application performance requirement must be guaranteed. At the same time, the affect on the application caused by the adaptation should be minimized, i.e., as smaller as better.

- The resource balance in the system should be considered.

Objective function

Considering the above prerequisites, we have the following three objectives for the optimization:

(1) Cost increase: the cost increase should be as small as possible under the condition that the performance of the application is guaranteed.

Assuming that the application k can be allocated the desired amount of resource r_{kd}, then the cost is

$$C_{kd} = p_k \bullet r_{kd} = \frac{1}{1 - \frac{R_{k-1,used}}{RC}} \bullet r_{kd} \qquad (5.2.1)$$

here p_k denotes the resource price in the system for the application k, it is calculated according to the total amount of resources already allocated to the k-1 applications in the system $R_{k-1,used}$, and the system resource capacity RC.

If r_{kd} cannot be satisfied by the system, but instead, r_k can be allocated to the application, the cost for the application becomes

$$C_k(r_k) = p_k \bullet r_k = \frac{1}{1 - \frac{R_{k-1,used}}{RC}} \bullet r_k \qquad (5.2.2)$$

Therefore, the cost increase function is

$$f_1(r_k) = \Delta C(r_k) = C_k(r_k) - C_{kd} \qquad (5.2.3)$$

(2) Efficacy loss: similar to the cost increase, the efficacy loss of the application should also be as small as possible under the prerequisite that the performance of the application can be satisfied.

Namely, when the resources assigned to the application change from r_{kd} to r_k, the efficacy loss is:

$$f_2(r_k) = El(r_k) = Ef(r_{kd}) - Ef(r_k) = Ef_{kd} - Ef(r_k) \qquad (5.2.4)$$

(3) Total resource deviation: in order to keep the overall system resources balanced, applications should be discouraged for using the resources which are relatively scarce. Instead they should be encouraged to use those resources which are relatively abundant in the system. This means, we try to allocate the resources to applications along the direction of the total system resources RC. In other words, Dev_{Rused}, the resource deviation of the total used resources $R_{k,used} = R_{k-1,used} + r_k$ should be as small as possible. Namely

$$f_3(r_k) = Dev_{Rused} = 1 - \cos(R_{k,used} \wedge RC) \qquad (5.2.5)$$

Among the above functions, $f_1(r_k)$ and $f_2(r_k)$ represent the application benefit. They reflect the performance satisfaction degree of an application. $f_3(r_k)$ represents the system benefit. It stresses the resource balance of the system. Hence, the goal is then to find values for r_k, so that all the three conditions can be satisfied as far as possible. Obviously this problem is a multi-objective function optimization problem (MOP). We use the weighted sum method to construct the evaluation function for solving this MOP problem. Thus, the relative importance of both the application and the system benefit can be controlled in the evaluation function. Considering the third objective function has a value between 0 and 1, whereas the first two objective functions have different meanings physically, and they may have different scale, we normalize the first two functions using the corresponding value at the point of ideal resource requirement. Hence, the evaluation function becomes:

$$F(\mathbf{r_k}) = \alpha_1|\Delta C(\mathbf{r_k})/C_{kd}| + \alpha_2|El(\mathbf{r_k})/Ef_{kd}| + \alpha_3|1 - \cos(\mathbf{R_{k,used}} \wedge \mathbf{RC})|,$$
$$\alpha_1, \alpha_2, \alpha_3 > 0 \qquad\qquad (5.2.6)$$

Here α_1, α_2 and α_3 are the weight parameters of the cost increase, efficacy loss and total resource deviation, respectively. They can be specified by the system to indicate the relative importance of application performance and the system resource balance and be provided by the application to convey which issues, cost increase or efficacy loss, are more important to the user. In order to keep the simple treatment, we suppose that the efficacy loss and the cost increase have the same importance to the application, namely $\alpha_1=\alpha_2$; and the ratio between α_1 (or α_2) and α_3, is specified by the node system administrator at the node initiation. After some experiments, we found that when $\alpha_1=\alpha_2=0.8$, and $\alpha_3=0.2$, both the application performance and the system resource balance can be well kept. Therefore, by default, we assume $\alpha_1=\alpha_2=0.8$, and $\alpha_3=0.2$, namely the application performance is more important than the system resource balance requirement.

Since the form of cost increase $\Delta C(\mathbf{r_k})$ and the total resource deviation are fixed, the complexity of the ultimate objective function depends in fact mainly on the efficacy function. If the efficacy function is linear or quadratic, the objective function is quadratic and the optimization problem becomes quadratic programming. To sustain the generality, we do not focus on solving the specific quadratic problem, whereas on solving the general nonlinear problem.

Constraints

The constraints of the optimization problem are constructed by the ARVS and the current available resources in the node system, namely

1. $\mathbf{r_k} \leq \mathbf{Rav_{k-1}}$

2. $\mathbf{r_k} \in ARVS$

3. $m_k \geq m_{k,used}$, i.e., for the application being adapted, the amount of memory resource should be greater than the amount already used by it. Note that if the application is the new arriving application, $m_{k,used}=0$, thus it does not need to be considered. To treat all cases in the same manner, we construct a new resource vector $\mathbf{rm_{k,used}}$ using m_k, whose coordinates are zero except for m_k.

Among these constraints, the second constraint has been partially reflected by the objective function $f_1(\mathbf{r_k})$ and $f_2(\mathbf{r_k})$, representing the cost increase and the efficacy loss. Hence, it can be simplified as :

$$\mathbf{RV_L} \leq \mathbf{r_k} \leq \mathbf{RV_U}$$

Considering the first and the third constraint, the above constraints can be changed to:

$$max\{RV_L, rm_{k,used}\} \leq r_k \leq min\{RV_U, Rav_{k-1}\}$$

Hence, the problem becomes a bound constrained nonlinear optimization problem.

5.2.4.2   Nonlinear Simplex Method

Various efficient optimization algorithms for solving the bound constrained nonlinear problems are available in the literature, including Barrier Newton Method, Bound Constrained Ellipsoid Method, Bound Constrained Newton, Nonlinear Interior-Point Method, NPSOL Wrapper, Parallel Direct Search (PDS) Method, Sequential Quadratic Programming (SQP) etc. The common characteristic of these algorithms is that all of them need the Gradient or the Gradient and Hessian of the objective function. For our optimization problem, the Gradient and Hessian of the objective function part $f_1(r_k)$, $f_2(r_k)$ and $f_3(r_k)$ are needed. For $f_1(r_k)$ and $f_3(r_k)$, i.e., the cost increase and the total resource deviation, the Gradient and Hessian can be calculated in the node system according to their definitions. However, the Gradient and Hessian of $f_2(r_k)$, i.e., the efficacy loss, have to be provided by the applications or be calculated by the system using numerical methods. This makes both the applications and the system more complicated.

The extended Simplex algorithm can also be used to solve the nonlinear, bound constraint problem, although it is a simple algorithm for solving the general optimization problems and not specialized at solving bound constraint nonlinear problems. The most important advantage of the Simplex method is that only values of the objective function are needed during the optimization procedure, neither the Gradient nor the Hessian of the objective function are required. Hence, we have selected the Simplex algorithm to solve our problem.

The basis of the Simplex algorithm was developed in 1962 [SHH62], but in 1965 Nelder and Mead are attributed with developing its modern form [NM65]. It is a robust method for optimization and is generally suitable for a wide variety of optimization problems. The Simplex algorithm finds solutions of any n-dimensional objective function through searching through the variable space. In a n-dimensional space, a $n+1$ geometric figure is called a simplex, whose corners are called vertices. Based on the initial $n+1$ trials of the values of the vertices, a search direction can be found, and a simplex can be constructed. Through evaluating values of the vertices of the new simplex, a new search direction is found, and a "better" simplex can be fixed. This procedure repeats until the optimum is found.

Figure 5.7 illustrates a simplex and its vertices in two dimensions. Let $f(x)$ denote the function to be minimized, where x is the n-dimensional vector of the conformational variables. The algorithm proceeds as follows:

1. Select a starting simplex represented by $n+1$ vertices.

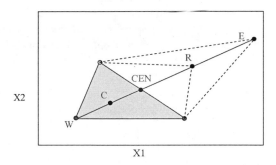

Figure 5.7: Illustration of simplex steps in two dimensions
W - worst (maximum) point, R - reflection point, E - expanded point,
C - contracted point. Grey area indicates the initial simplex

2. Evaluate the target function f at the $n+1$ vertices. Select the worst point x_h such that $f(x_i) \leq f(x_h)$ for $i=1, 2,...,n+1$, and the best point x_l such that $f(x_l) \leq f(x_i)$ for $i=1, 2, ..., n+1$. Calculate the centroid x_0 of the simplex by $x_0 = \sum x_j/n$, where the sum is for all j, $j \neq h$, i.e., the centroid excludes the worst point x_h.

3. Calculate the reflection point $x_r=x_0-a(x_h-x_0)$, where $a > 0$ is the reflection coefficient, and evaluate the function f at x_r. If $f(x_r)$ lies between $f(x_h)$ and $f(x_l)$, the reflected point is kept, i.e., x_r replaces x_h.

4. If $f(x_r) < f(x_l)$, further improvement can be expected by continuing along the same direction as the entire reflection (x_0 to x_r). Thus, an expansion is performed to a new point $x_e=x_0 + g(x_r - x_0)$, where $g > 1$ is the expansion coefficient. If $f(x_e) \leq f(x_l)$, the expanded point is kept, i.e., x_e replaces x_h , Otherwise, the original point x_r replaces x_h.

5. If the reflection process gives a point x_r for which $f(x_r) > f(x_i)$ for all i except $i=h$, and $f(x_r) < f(x_h)$, x_h is set equal to x_r, and a contraction operation is performed to find a new point $x_c=x_0+b(x_h-x_0)$, where $0 \leq b \leq 1$ is the contraction coefficient.

6. if $f(x_r) > f(x_h)$, x_h is left unchanged but the contraction operation is still performed. If $f(x_c) < min[f(x_h), f(x_r)]$, the contracted point is retained, i.e., x_h is replaced by x_c . If $f(x_c) \geq min[f(x_h), f(x_r)]$, the entire simplex is contracted around the best point by replacing all x_i by $(x_i + x_l)/2$, and the reflection process is restarted.

7. The new simplex is checked for convergence. The algorithm is assumed to have converged when the standard deviation Q of the function at the $n+1$ vertices of the current simplex has decreased past some small tolerance value, i.e., $Q=\sum\{[f(x_i) - f(x_0)]^2/(n + 1)\}^{1/2} \leq \epsilon$. If this inequality is not satisfied, then return to step 2, and perform a new series of simplex moves starting from the current simplex.

We have implemented the above Simplex algorithm in our AN node system. For our problem, the objective function is a 3-dimensional function, since the network bandwidth, computational and memory resource are considered. The complexity of the objective function varies with the efficacy function of applications, because the cost function and the total resource deviation have fixed forms according to their definitions. In addition, in order to control the time overhead introduced by the admission control procedure, we limit the iteration number of the Simplex algorithm. I.e., we add another termination criterion in step 7. If the iteration number exceeds the specified value, the iteration ends. In this case, we assume that no solution is found.

In conclusion, the stability and the lack of usage of derivatives make the Simplex method especially helpful for our resource adaptive admission control problem. However, since the Simplex algorithm uses a "geometric" method to achieve function minimization and employs relative arbitrary values for the deterministic factors that describe the "movement" of the simplex in the merit space, it is relative slow in converging to local minimum. Theoretically it is applicable to small scale optimization problems. A previous general rule of thumb is that the dimensionality should be limited to around 10 to 15 variables. This exerts of course no limitation to our problem, as our problem has only 3 variables; it is a quite small optimization problem. Therefore, we think the convergence speed does not play an important role for our optimization problem. We will give some evaluation about that in the following section.

5.3   Evaluation

The adaptive admission control mechanism is outlined as follows:

1. Applications carry their desired resource requirement and ARVS information in their first packets to the AN nodes.

2. Whenever there is a new application request at a node system, the node system checks the current system resource status, in order to decide whether there are enough resources for the new application.

3. If it is found that a certain resource type cannot meet the requirement of the application, the adaptation begins. The adaptation begins first within the new application. If it fails, a helpful application is searched from the table CommonARVS, and the adaptation continues within this application.

4. If an adaptation is successful and it is decided that the application can be accepted into the system, the ARVS information of the application is inserted into the CommonARVS table in the proper position, using the binary search method.

From the above key actions occurring during one admission control procedure, we can see that the overhead introduced by the adaptive admission control mechanism involves two aspects: transmission bandwidth overhead and processing overhead. The transmission overhead is caused by carrying resource requirement and adaptability information in the packets to the AN nodes; and the processing overhead, i.e., the extra processing time spent on admission control, comprises mainly three parts: time for system resource status checking, time for adaptation, including optimizations and searching for candidate applications, and time used for ARVS registration. System resource status checking is implemented through invoking a function provided by the NodeOS. As introduced in chapter 6, the function is implemented through reading the values in the resource controllers allocated to each application in the system. The search for the candidate applications means in fact only fetching an application from the top of the CommonARVS table, since applications have already been sorted in descending order according to the excess of efficacy and cost when they are admitted to the system. Therefore, these two parts can be neglected. In the following, we evaluate the processing overhead introduced by the adaptation and registration procedure, and the bandwidth overhead caused by the transmission of the adaptability information.

5.3.1 Adaptation

5.3.1.1 Optimization Time

As introduced in section 5.2.3, the adaptation procedure includes 3 steps. During the first step, an optimization method is used to find a new value of RV for the application. Therefore, the first step is in fact an optimization procedure. During the second and third step, first a possible helpful application is selected, and then it is changed to an optimization procedure. Hence, from estimating the time spent on each optimization, we can evaluate the maximum processing cost of an adaptation.

From (5.2.6) we can see that the final objective function depends on the system resource capacity $\mathbf{RC}$, the available resources at one time point in the system $\mathbf{R_{av}}$, the desired resource requirement r_{kd} as well as the efficacy function of the application. When the former three numerical values are determined, the objective function depends in fact only on the efficacy function of the application. Therefore, we study the time spent on the optimization under different complexities of efficacy functions. The following tests are performed in a computer with mobile CPU 866MHz under windows XP system.

In the first test, we suppose a linear efficacy function, namely

$$Ef = \frac{0.58}{20000}b + \frac{0.18}{300000}c + \frac{0.24}{40000}m$$

where (b, c, m) denotes the network bandwidth, CPU and memory resource respectively. In this test, the number of iterations is 67, and the time spent is 1.0144ms. Note that here, the weight parameters for the cost increase, the efficacy loss and the resource deviation is under the default value, namely $\alpha_1 = \alpha_2 = 0.8$, and $\alpha_3 = 0.2$. In addition, the available resource in the system is $\mathbf{Rav} = (18000.0, 2000000.0, 6959260.0)$, the system resource capacity

RC=(1.0E7, 3.0E7, 1.0E8), the resource requirement r_{kd}=(20000.0, 300000.0, 400000.0), and RV$_L$=(12000.0, 130000.0,300000.0), RV$_U$=(40000.0, 500000.0, 600000.0).

In order to give a general overview about the iteration number and duration under the linear efficacy function, we generate the coefficients of the linear efficacy function randomly and repeat the optimizations 100 times under the same resource conditions. Figure 5.8(a) shows the result. The link diagram illustrates the distribution of the iteration number. And the right diagram shows the duration correspondent with the iteration number, it is the average of the time cost by the optimizations when their iteration numbers fall in the corresponding scope. From this figure we can see that the iteration number between 80 and 90 has the highest possibility (23%), and the corresponding time cost is 1.9ms. In addition, the iteration number for most optimizations (99%) is below 200; therefore, we limit the maximum iteration number of our optimization algorithm to 200. That means, if the iteration number exceeds 200, we assume that the optimization fails.

In the second test, we select a quadratic efficacy function with the following form:

$$Ef = A_1b^2 + A_2c^2 + A_3m^2 + A_4b + A_5c + A_6m$$

$A_1, A_2, A_3, A_4, A_5, A_6$ are positive coefficients. Note that here only a specific form of quadratic efficacy function is given. We have not used the terms like bc or cm, because we do not see application scenarios for that and did not want to make the study unnecessary complex.

Similar to the first test, we have generated the coefficients randomly and repeated the optimization 100 times under the same resource condition as the first test. Figure 5.8(b) illustrates the result for quadratic efficacy function.

We have also performed the test using the following cubic efficacy function, and the result is shown in Figure 5.8(c).

$$Ef = A_1b^3 + A_2c^3 + A_3m^3 + A_4b + A_5c + A_6m$$

From Figure 5.8 we can see that for the optimizations with the same iteration numbers, those with the cubic efficacy function cost the longest time and those with the linear efficacy function cost the shortest time. However the difference among the time cost is not obvious. In addition, the average iteration numbers of optimization with quadratic and cubic are even smaller than that with linear efficacy.

The reason is that the convergence speed, i.e., the iteration numbers, of the Simplex algorithm is mainly determined by two factors, one is the number of the variables of the objective function, and the other is the features, such as the geometric form of the objective function. When the number of the variables of the objective function increases, the number of vertices needed to be evaluated increases, and the search space of the simplex increases also. As a result, both the iteration numbers and the time spent on

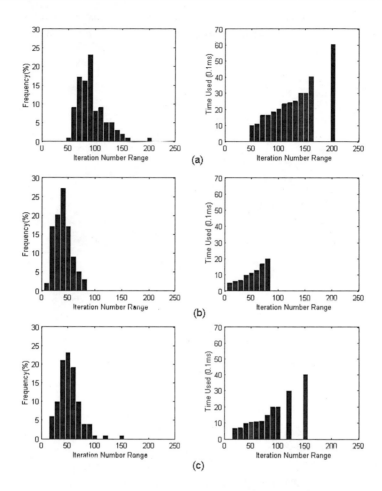

Figure 5.8: Efficacy function and optimization time
(a) linear efficacy function (b) quadratic efficacy function (c) cubic efficacy function

each iteration may increase. In addition, since the Simplex algorithm seeks the local optimum, when local minimal values exist, the algorithm can find a solution fast. Figure 5.9 illustrates the objective function with linear and cubic efficacy function in our example. We can see that there exist some local minimums for the objective function with cubic efficacy function. Therefore, its average iteration number is smaller than that with a linear efficacy function. It is similar for the objective function with a quadratic efficacy function.

On the other side, the increase of the complexity of the objective function affects only the compute time in each vertex evaluation. Unless the objective function is very complex from the perspective of computing demands, there is not great difference in the time cost among different objective functions in case the iteration number and the variable number are the same.

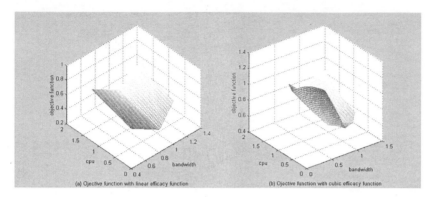

(a) Ojective function with linear efficacy function   (b) Ojective function with cubic efficacy function

Figure 5.9: Objective function

So far, we considered only the three major types of resources in the system; therefore, we think Figure 5.8 can roughly represent the time spent on one adaptation within an application with polynomial efficacy functions. We believe that also in the near feature, the number of resources types concerned in AN nodes will still meet the general rule of thumb of the Simplex algorithm that the number of variables should be limited to 10 to 15.

For comparison purposes and further study the convergence speed of the Simplex method when it is applied in our resource adaptation, we have tried to solve our problem with the SQP optimization method.

The SQP method [Fle86] [Spe98a] is regarded as one of the most efficient methods for solving NLP problems. The basic idea of the SQP method is to replace the original NLP problem by a sequence of quadratic problems (QPs). Each QP has linear constraints and a quadratic objective, and is easily solvable.

Considering the standard constrained nonlinear problem (5.1.4) to (5.1.6), the general idea of the SQP method is summarized as follows:

First, the SQP method replaces the objective function with the quadratic approximation:

$$minimize \quad \frac{1}{2}p_k^T \nabla^2 L(x_k)p_k + \nabla f(x_k)^T p_k + L(x_k)$$

where the $L(x)$ is the Lagrangian function, $L(x, \lambda, \mu) = f(x) - \sum \lambda_i g_i(x) - \sum \mu_i h_i(x)$

Second, the constraints are replaced by linear approximations, namely

$$subject \ to \quad g_i(x_k) + \nabla g_i(x_k)^T p_k = 0$$
$$h_i(x_k) + \nabla h_i(x_k)^T p_k \geq 0$$

That is to say, at each major iteration, the SQP method obtains an updated search direction by solving a QP sub-problem. The objective of the QP sub-problem is a quadratic approximation of a modified Lagrangian function that depends on the nonlinear problem objective and constraints. The constraints of the QP sub-problem are linearization at the current point of the nonlinear problem constraints.

DONLP2 [DON] is a free software package using SQP method for solving general nonlinear problems with mixed equality and inequality constraints. Bound constraints are integrated in the inequality constraints. It uses a sequential equality constrained quadratic programming method, which is considered as one of the most efficient methods for solving nonlinear problems of small to medium size. It employs a slightly modified version of the Pantoja-Mayne update for the Hessian of the Lagrangian, bounds on the variables are treated in a gradient projection like fashion. Details can be found in [Spe98a] and [Spe98b].

We have used DONLP2 to test our adaptation problem under the same resource condition as using the Simplex method. The average iteration number of 10 tests under the linear, quadratic and cubic efficacy function is 13, 6, 7 respectively, and the corresponding average time cost is 1.276ms, 3.734ms and 3.967ms respectively. Compared with the Simplex algorithm, the convergence speed of the SQP method is obviously better. However, the comprehensive time difference is not large; for the quadratic and cubic efficacy function, the cost is even a little bigger. The reason is that, if the efficacy function is linear, the Gradient and Hessian of the objective function are very simple, and the SQP method is fast. Otherwise, SQP has to take some time to solve the Gradient and Hessian numerically. Since our resource adaptation problem does not contain many variables, and the Simplex method does not need to know the Gradient or Hessian of the efficacy function from the applications, and avoids the problems of non-convergence caused by calculating the Gradient or Hessian of objective functions using the numerical method, we have selected the Simplex method in our system.

So far, we have only evaluated the time needs for an optimization procedure. Since the adaptation depends closely on the system resource status and the ARVS of each application in the system, we cannot make a statistic about how many adaptations occur in the second step.

5.3.2 Registration

As introduced in section 5.2.3, in the table CommonARVS, the ARVSs of the applications in the system are sorted in descending order on the excess S of the efficacy and cost, namely the sum of $Ef(r_{k'})-WE_{k'}$ and $HC_{k'}-C(r_{k'})$. During admission control, if no adaptation is needed or the adaptation within the new application succeeds, the ARVS of the application is inserted into the CommonARVS table using the binary search method. In case an adaptation among applications occurs and succeeds, the ARVS of the helpful application is re-ordered into the suitable position and the ARVS of the new arriving application is inserted into the table using the binary search method. Therefore, there are maximal two binary search procedures during one admission control procedure.

In case there exist k-1 applications in the system, the location searching steps are $log_2(k$-1$)$, i.e., there are $log_2(k$-1$)$ comparison computations. Considering that there are at least 4 times of comparison computation in each iteration during the optimization, for an unsuccessful adaptation with maximum 200 times of iteration, only when the number of the applications in the system is equal to 6.7×10^{240}, the cost for registration is roughly equal to an optimization procedure. Furthermore, during each iteration, different values of the objective function are compared, and there are also some other computations, e.g., the calculation of the centroid of a simplex etc; whereas in the registration, only the excess value S is compared. Therefore, practically the time spent on the optimization is much higher than that on the registration. Compared with the adaptation cost, the registration cost can be neglected.

5.3.3   Bandwidth overhead

Apparently, the transmission of the resource requirement and the adaptability information increases the length of the packets. From section 5.2.2, we can see that the resource requirement r_d causes an increase of 12 bytes, since each RV has a fixed length. However, because the length of both the efficacy function and the ARVS are variable, the bandwidth overhead caused by the adaptability information is not fixed; it depends on the complexity of the ARVS information of each application. For example, for a linear efficacy function with 3 terms, in case both the HC and WE are not omitted, the length of the ARVS is 50, therefore, together with the overhead introduced by the type and length indication of the RV and ARVS option field as well as 2 bytes zero filling for ARVS option payload, the total increased length is 72 bytes.

5.4   Discussions

First, so far we assume that each application knows its desired resource requirement, which may not be practical in some cases. In fact, the desired resource requirement can also be calculated according to the QoS requirement or the efficacy function and the cost function of the applications. This procedure can be implemented in the AN node systems, e.g., adding a software module in the admission control algorithms. It can also be implemented by the application itself, e.g., using a general agent.

Second, we have assumed in our adaptive admission control mechanism to carry the resource requirement and ARVS information in the first packets of applications from users to the network nodes. In the IP-based networks, not all the packets of an application pass through the same route from the source to the destination. However, this does not affect our mechanism. As mentioned in chapter 2, an active network can be considered as an overlay network on top of the traditional IP-networks: some nodes support application-specific computations, and are therefore called AN nodes. In case a user wants a specific computation in an AN node, he will first program the AN node and then send packets through this node to a destination. Although the packets may arrive at the destination through different "passive" nodes, they will pass the AN node. And for the AN node, the first packet is the packet that "programs" the node. In fact, currently the AN nodes usually use their own routing table to forward active packets. I.e., the EE owns a routing table, which can be configured by the System and Management EE. The NodeOS creates channels and forwards active packets according to the parameters denoting the next AN node provided by the EE. Thus, all the active packets can be controlled to pass the specific AN nodes, despite of the "low level" routing table at each node. Now in the AN test network ABone [BBR00], the routing tables in an AN node can be configured by the Anetd [DGM+01]. In the future, a mechanism (like the Path message in RSVP) may be developed to specify the nodes which should be passed through.

Third, global or local optimum is a discussion issue for optimization algorithms in general. Some conditions can be used to determine if a problem has a global optimum and if a solution is the global optimum for some problems. For our problem, this question is of minor concern, because our goal is to perform resource adaptation, and our task is to find a feasible solution with the help of the optimization method. We care only whether the solution is inside the ARVS, but it is less important whether the solution is the best one in the whole ARVS.

5.5   Related Work

RSVP [ZBH+97] employs admission control during the resource reservation setup. It uses some separate control path messages to carry the desired QoS information and reservation results. Unlike this scheme, our admission control mechanism adopts the in-band method to convey the application-specific resource adaptability information and the results of the admission control. In addition, in RSVP the admission control procedure is triggered by the reservation message originated from the receiver, but our admission control procedure is triggered by the "sender". Moreover, our admission control method checks also multiple types of resources in the system, and implements also an adaptation algorithm during the admission control.

[YG00] has introduced a resource trading model for active applications, which uses agents to exchange information between AN nodes and user applications. In this model, two types of agents communicate to seek an equilibrium. Each agent tries to optimize its own benefits: on the one side, the network system resource manager agents have the goal of maximizing resource usage while maintaining a good performance level. On the other side, user agents try to obtain a better quality-price relation for the resources consumed, and to efficiently manage their own budgets avoiding waste. Both types of agents are implemented as AAs with different privileges, and they communicate such that the resource managers can "sell" resources to the user agents at a price that varies as a function of the demand of the resource.

To some extent, our work shares the same ideas with this work: both of them use the resource price to reflect the resource status in the AN nodes. In addition, [YG00] involves also a tradeoff between buffers and network bandwidth. However, there are some essential differences between them. [YG00] is based on a generic communication abstraction between agents. The resource status information is exchanged between user agents and node system resource manager agents through special secure communication. The applications can adjust themselves according to the acquired resource status information in the networks. Therefore, [YG00] performs in fact application-side adaptations. Our approach adopts the system-side adaptation. There is no need to exchange the resource status information between users and AN nodes. Hence, our approach can adapt faster and reduce the transmission overhead.

[NL00] suggests a cost model for active networks. It has involved the relationship between network bandwidth and CPU processing resources from a higher level. It believes that flows in active networks are no longer of constant rate throughout the route from origin to destination due to the introduction of computation in network nodes. Therefore, the CPU (processing) resource cost can be calculated using a function of the ratio of the output rate to the input rate. It decides whether a processing at a node is cost-effective through defining and comparing two types of total cost, which consist of the processing cost and transmission cost per unit of bandwidth. Both [NL00] and our work advocate that CPU resource consumed by active applications should be taken into account by AN nodes, and involve the tradeoff between the CPU and bandwidth resource. However, [NL00] abstracts and charges the CPU consumption of an application according to the variations of the network bandwidth at output and input side, and concerns the resource tradeoff by deciding in which network node an agent should be launched through comparing the CPU costs at different nodes. However, we believe that this approach for calculating CPU costs is unrealistic, since the "compression" ratio is application and algorithm specific and not related to CPU demand.

Czajkowski et. al. have discussed the resource consumption tradeoffs from the perspective of servlets in extensible Internet servers [CCH+98]. They have analyzed the advantages of being resource aware for servlets and advocated providing resource availability feedback for servlets. Based on the J-kernel [HCC+98], the servlets can acquire some information about resource availability in the servers and adapt their behavior in response to the resource load fluctuations. However, they have not answered the question of how to determine which tradeoffs to pursue when there are several possibilities from the point of view of resource availability. We deal with this problem by examining the system resource availability using the concept of resource price. In addition, our work concentrates on system-side adaptations as well as how to express the resource adaptability of applications.

[Khan98] presents a utility model for an adaptive multimedia system (AMS) with multiple concurrent sessions, where the quality of individual sessions can be dynamically adapted. In this model, each session has a quality profile consisting of a set of discrete operating qualities. Any operating quality can be converted to the required resources using a quality-resource mapping and to a session utility using a quality-utility mapping. The main problem in a multisession AMS is to find an operating quality for each session so that the overall system utility is maximized under the resource constraints. This problem is then formulated as the multiple-choice multi-dimension 0-1 knapsack problem.

To some extent, the sessions in the utility model can be considered as applications in our system. Therefore, the utility model and our work have some similarities: both view the adaptation as an optimization problem. However, the two models are fundamentally different in their problem formulation, scope and objective. The utility model uses a set of discrete operating qualities to describe the adaptability of each session, and assumes that each operating quality can be mapped uniquely to the required resources. However, we use the ARVS to express the adaptability of an application, which indicates implicitly that the quality-resource mapping is not unique, different combinations of resources can sometimes result in the same performance.

In addition, the utility model formulates the adaptive multimedia system as a multi-choice, multi-dimension 0-1 knapsack problem. Namely an operating quality is selected for each session so that the system utility (i.e., the sum of the session-utilities) is maximized. This is an integrated optimization problem. But our adaptation problem is actually formulated as a bound-constraint continuous optimization problem. The minimal value of the objective function is solved during the optimization procedure. Moreover, the admission control in the utility model has different meanings compared with that in our system. Besides finding a feasible solution so that the $n+1$ sessions can share the currently available resources, the system utility under $n+1$ sessions U_{n+1} must be greater than that under n sessions U_n, namely $U_{n+1} > U_n$. Otherwise, the session should be rejected. However, besides stressing the application performance, our admission control algorithm emphasizes the system resource balance and the total system resource utilization. To improve the utilization of the different kinds of resources in the system and keep them in balance belong to the most important goals of our admission control. Finally, [Khan98] gives only an abstract utility model, the quality-resource mapping and quality-utility mapping in the prototype are only randomly generated tables; resources have no concrete meaning, and the system utility is evaluated as a performance index. In short words, the utility model involves in fact a method for gracefully degrading the QoS of multimedia sessions by enlarging the total system utility. Whereas our system concentrates on adaptation among different types of resources, and keeping them in balance.

[HWD98] suggests a system that enables criticality-and QoS-based resource negotiation and adaptation for mission-critical multimedia applications. It introduces a two-phase QoS adjustment scheme for allocating resources for a new stream. The first phase is called the shrinking phase. It reduces the QoS of the executing streams to accommodate a new stream, achieving the goal of maximizing the number of concurrent streams. The second phase is called the expansion phase. It expands the QoS of the concurrent streams once the new stream is admitted, achieving the goal of QoS maximization. In the second phase, a criticality-based multiresource preemption scheme is employed in case of resource contention when the system has not sufficient resources to meet the minimum QoS requests. Two approximation algorithms are developed toward the goal of supporting high-criticality applications and maximization of the total number of concurrent applications.

Both [HWD98] and our work have introduced an optimization method with multiple objectives for adaptation in the networks. The objectives of the optimization in [HWD98] include to meet the timing constraint of multimedia applications, maximize application QoS and the number of executing high-criticality applications. Whereas the objectives of our optimization are to satisfy the application performance requirement, keep the system resource balanced and improve the utilization of different types of resources. However, [HWD98] implements the optimization in two phases. Moreover, during maximizing the number of the concurrent streams, the problem is converted to an integrated optimization problem. In order to avoid the NP-hard optimization problem, an approximation solution that is near optimal is adopted. But we use the continuous optimization method to formulate and solve our adaptation problem, and reach the multiple objectives simultaneously.

5.6 Summary

This chapter has presented an adaptive admission control mechanism in AN nodes based on the study of the adaptability of active applications. Applications carry their resource requirement and adaptability information to the networks together with the execution request. Adaptations may occur during the resource admission control procedure. During the admission control phase, not only the application performance is taken into account, but also the system resource balance and the total resource utilization are considered. The result of our admission control is in fact a tradeoff between the performance of a single application and the overall system resource utilization. In addition, the adaptive admission control mechanism realizes also the application-specific adaption in the networks, which avoids the complexity and transmission overhead caused by integrating the adaptation mechanism in applications and the information exchange between applications and networks.

We have also evaluated the adaptive admission control mechanism in terms of the extra processing time introduced by the optimization algorithm and the bandwidth overhead introduced by the transmission of the application resource requirement and the adaptability information.

Chapter 6

System Implementation

This chapter presents the implementation of the AN node system realizing the resource adaptation algorithm suggested in the previous chapters. First the architecture of the node system is introduced, focusing on the various components that provide the AN environment and the resource management functions. Then the resource allocation and control scheme supporting the resource adaptation algorithm is addressed, including the resource accounting and limiting techniques of the diverse resources used by applications. Following this the performance evaluation concerning the implementation is presented. Finally, the related work with regard to the resource accounting and control in Java language is discussed.

6.1 Active Node Architecture with Resource Management Subsystem

The AN node architecture which implements our adaptive resource admission control mechanism is illustrated in Figure 6.1. Generally, it consists of the three major components suggested by the active network working group in [Cal99], namely the active application (AA), the execution environment (EE) and the node operating system (NodeOS). Through these components, the general function of an active node, namely the environment for supporting the loading and execution of the programs injected by applications and some basic functions and services that can be used by applications, is provided. In addition, a resource management subsystem is integrated in the NodeOS to carry out the adaptive resource admission control mechanism suggested in the previous chapters. Currently, the whole architecture runs as a process under the Windows XP system, denoted as AN-withARM (Active Node architecture with Adaptive Resource Management subsystem). In the following sections, we explain these components in detail, concentrating on their structure, function and basic features, as well as how the resource management functions are provided. We begin with the NodeOS at the lowest layer.

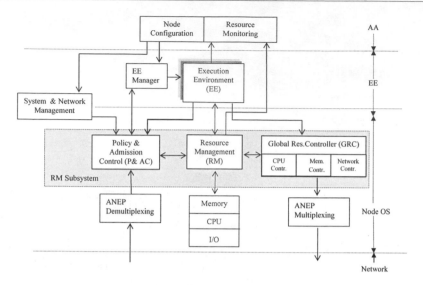

Figure 6.1: The node architecture with resource management subsystem

6.1.1   Node Operating System (NodeOS)

As introduced in chapter 2, basically the NodeOS is responsible for managing the resources in the node system and providing some basic functions that can be invoked by active applications through EE. Therefore, our adaptive admission control mechanism is implemented in this part. To reduce the work for developing a NodeOS by implementing the NodeOS APIs specified in [AN01], we have selected the Janos [Janos] Java NodeOS (Jnodeos) as the basis of our NodeOS among the several approaches introduced in section 2.4. The most important reason is that it implements the NodeOS API in Java, supports a Java-based EE, and can execute on a regular Java Virtual Machine (JVM). In addition, Jnodeos uses also some resource management abstraction functions, which is helpful for us to control the resource usage in the node system to realize the adaptive resource management mechanism.

6.1.1.1   Janos Java NodeOS (Jnodeos)

Jnodeos is a part of the Janos (Java-oriented Active Network Operating System) project at the University of Utah, which aims at developing an operating system for AN nodes, preventing separate active applications from interfering with each other and providing node administrators with controls over the resource usage of applications [THL01]. Jnodeos implements the NodeOS API specification in Java, therefore may support various Java-based EEs, e.g., ANTS2.0, to run on top of it. Jnodeos supports all of the major NodeOS abstractions such as Threads, Channels, Domains, ThreadGroups, and DemultiplexKeys that encapsulate the diverse system resources.

In addition, Jnodeos contains a JanosVM emulation library, which provides JanosVM facilities on standard JVMs. JanosVM [PGH+01] [THL01] is an extended, resource-aware Java Virtual Machine (JVM), directly based on the KaffeOS [BHL00] implementation. It is responsible for providing independence between concurrently executing Java applications. The JanosVM-based functions, such as limiting the threads created by an AA in one domain, and reclaiming resources from terminated AAs, provide a basis for us to integrate the main resource accounting and controlling functions needed to realize the adaptive resource admission control into the NodeOS.

Figure 6.2 illustrates the relationship between the functional components in our node architecture. With the help of the JanosVM emulation, Jnodeos runs on standard JDK 1.1-compliant runtimes. And the ANTS2.0, a Java-based EE that we use, can run directly on top of Jnodeos.

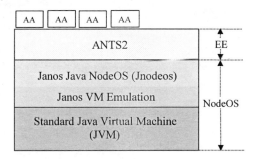

Figure 6.2: Components used

6.1.1.2   Main changes to Jnodeos

In order to be able to control the amount of CPU, bandwidth and memory resource used by each active application within the agreed value from the admission control, which may be changed during the execution of the applications due to the adaptations, we have made some changes and added some new functions to the Jnodeos.

The most important change to Jnodeos is that we have integrated a resource management subsystem (RM subsystem) into it, to perform the admission control, resource adaptation, accounting and scheduling etc. We will discuss this subsystem in detail in the following section. Due to the introduction of the RM subsystem, some changes must correspondingly be made to the Jnodeos, for instance,

- Some functions are extended and re-organized into the RM subsystem. E.g., a resource counter has been added to each application domain; some thread pools are re-arranged to the scheduler module described in the next section; and some methods are added to process the exceptions etc;

- Some data structures in Jnodeos have been changed, e.g., new fields are added to record the intermediate results or to process the new functions;

- Some changes have also been done related to the packet demultiplexing and multiplexing functions provided by Jnodeos. Since we have made some changes in the packet header, such as adding the desired resource requirement and adaptation capability information, and increasing the values of the Type ID that can be processed by the node system (see section 6.1.2), the corresponding information must be identified. Particularly, the check of the Type ID functionally belonging to the Policy and Admission Control module is integrated into the demultiplexing functions. In case packets need to be sent, this information should also be processed in the multiplexing phase.

Furthermore, some interfaces to other modules for the purpose of communication have been added to Jnodeos. This covers mainly three aspects:

- The EE manager in the EE layer needs information from the underlying NodeOS to decide whether new EE instances can be created considering the resource status in the node system.

- The NodeOS needs also to establish contacts with the upper layers for the configuration purpose.

- The NodeOS is responsible for providing the states of different kinds of resources in the node system for the upper layers.

Another change to Jnodeos is caused by the introduction of the new resource description method in our node architecture. In Jnodeos, resources are specified in a precise, hardware-dependent way, for example, physical memory limits are specified in terms of memory pages. Since we have introduced RV to describe the resource usage in the system, we add a method for the mapping between the RV used in our RM subsystem and the precise values needed by Jnodeos. For this purpose, we have added the classes of CPUSpec and NetBandwidthSpec etc. that were not provided by Jnodeos and implemented the methods such as RVtoMemSpec, RVtoCPUSpec as well as RVtoNetBandwidthSpec etc. to perform the conversion.

6.1.1.3   Resource management subsystem (RM subsystem)

This is the main block we have implemented in the NodeOS to manage the resource usage in the node system in order to realize the suggested adaptive admission control mechanism. As shown in Figure 6.1, functionally the RM subsystem consists of mainly 3 modules: policy and admission control (P&AC), resource management (RM) and global resource controller (GRC). It is responsible for performing the resource admission control, adaptation, accounting, and scheduling. In addition, some query functions about various resource states in the system are also implemented in the RM subsystem. In the following, we introduce the main functions implemented in each module, the concrete resource accounting and control scheme will be introduced in section 6.2.

Policy & Admission Control (P&AC)

Policy control checks whether packets from end-users are permitted to enter the node system and distinguishes the priorities of each end-user process in the node system. So far, following aspects related to the policy control are implemented in our node system:

1.   The Type ID in the ANEP header is used to determine whether packets from active applications are permitted to enter the system and can be processed. As it will be introduced in section 6.1.2, two kinds of packets are authorized to run in our node architecture using the Type ID field in the packet header according to the two different code distribution mechanisms, namely the on-demand code loading and the associated code loading. Although the Type ID field in the ANEP header is originally designed for distinguishing the types of EEs for processing the packets specified by the applications, we believe that it is a simple and fast way to perform the policy control to reduce the amount of packets into the system, as long as not too many types of EEs are supported simultaneously in one AN node system, which is true at the moment and in the near future. Since the Type ID is carried in the normal header of a packet, as shown in Figure 6.5, this function is indeed part of the demultiplexing phase, which distinguishes the protocols that a packet uses, e.g., IP/UDP/ANEP or IP/TCP/ANEP etc.

2.   We have assigned some applications, such as RemoteMonitor[1], the privilege to invoke directly the functions and services provided by the NodeOS, e.g., the resource status query functions, instead of using the interfaces provided by the EEs. Through this way, some processing overhead can be avoided. This function is implemented through retrieving the Protocol Identifier (Protocol ID) and the Method Identifier (Method ID) carried in the packets from users. Both the Method ID and the Protocol ID are compared with that stored in the system. If both are matched, a sign is marked for the corresponding EE instance created for this application. Then during the execution of the application, the functions and services provided by the NodeOS are allowed to be invoked directly.

3.   Two levels of priorities are assigned to the applications running in the node system. In order to be able to take advantage of the resource adaptation function provided by the node system, applications should specify their desired resource requirement r_d and the adaptation capability ARVS in the packet header, as described in section 6.1.2. However, to provide a downward compatibility, applications not specifying the required resource r_d and ARVS should also be accepted by the AN nodes. Therefore, in our implementation, we distinguish both kinds of applications in the system by assigning different priorities to

---

[1]We have developed this application on top of our node system to monitor the resource status at a remote active node. More information about it can be found in chapter 7 and [LWX03].

them. Those applications not specifying r_d and ARVS have lower priority; they can only be accepted when there are available resources in the system, no adaptations will occur for them. During the execution, threads of these applications have always the lowest priority, and their performance can not be guaranteed.

4. The identical policy is applied to the different types of resources in the node system. Since we use RVs to denote the resources in the node system, each type of resource has the same position in the system. E.g., threads of applications not specifying the r_d and ARVS have the lowest priority for consuming all types of resources.

All the parameters related to these policies, including the Type IDs permitted by the system, the Protocol IDs and Method IDs with the privilege to invoke the functions and services provided by the NodeOS directly, as well as the priorities of the applications with or without desired resource requirement r_d and ARVS, are configured through the System and Network Management Module.

Admission control determines whether an application can be executed in the node system at a certain time point. In other words, it decides whether there are enough resources in the system for creating a new EE instance and run the application-specified program under the specified resource requirement. The suggested resource adaptation algorithm is implemented in this module. During the admission control, the RM subsystem attempts to bind the system resources to the applications. The information needed for performing the admission control is acquired from the RM module. Applications are executed in the node system under the control of the GRC according to the result of the admission control.

In conclusion, the procedure and main tasks performed in the P&AC module are to:

- Check whether a packet is permitted to enter the node system according to the Type ID.

- Ask the RM module about the resource status in the system, including the available resources and the CommonARVS table recording the information about the candidates for adaptation among applications.

- Initiate the adaptive admission control procedure, during which adaptations within and among applications may happen, if the r_d and ARVS are specified in the packets.

- Check and mark the privilege and priorities of the application in the newly created EE instance.

- Inform the result, i.e., the agreed amount of the resources that the new EE instances can consume, to the RM module if a positive decision is made, where the information is stored and used by the GRC module to control the execution of the EE instance.

- Notify the EE Manager module that a new instance should be created with the given amount of resources.

- Prepare a negative response to the end-user application in case a negative decision is made during the admission control.

Global Resource Controller (GRC)

The global resource controller (GRC) schedules the execution of the end-user applications and the transmission of packets to the next node, monitors the resources consumed by each EE instance on behalf of applications, and controls them to be within the agreed value. In our node system, when an EE instance is created due to the arrival of a new application, a thread pool is created for it. Several parameters are specified when creating the thread pool, including the maximum number of threads in the pool, the resource controller to be used and the stack size for each thread. In addition, the thread pool is also pre-allocated a definite number of idle threads. These idle threads are dispatched to process events such as timeout, exceptions and receiving packet from channels. Therefore, the resources that an application needs are mapped into the hybrid thread/event and channel model in the node system. By scheduling the execution of the threads and channels, the applications accomplish their tasks, consuming the allocated resources. Generally, we adopt the pre-accounting technique in order to avoid the resource overuse. Namely before the agreed resource limit to an application is exceeded, actions are taken to prevent the possible resource overuse. However, due to the characteristics of each type of resource, the implementation of the controllers of different types of resources is different. In the following, we give an overview about the resource controllers, detail information about their implementation will be presented in section 6.2.

CPU Controller   When an EE instance is initiated due to the arrival of a new application, a thread pool is created and a CPU controller is allocated for it. All the threads used for packet receiving, sending and processing are spawned from this pool. The CPU controller counts the CPU resource used by all the threads in the thread pool, and controls it to be within the agreed value determined in the admission control phase. In case adaptations concerning the EE instance occur, the CPU controller controls the execution of the threads according to the new value.

The CPU controller assigns priorities to the threads according to the amount of the resources that the EE instance can acquire, and uses the priority-based scheduling algorithm to schedule the threads in the thread pools. For the threads with the same priorities, the round-robin algorithm is used. The CPU controller counts the number of the bytecode instructions executed by all the threads on behalf of the EE instance in a certain time interval. Through setting the value of the accounting time interval, the pre-accounting technique can be realized. When the CPU controller detects that the CPU limit will be broken, actions such as decreasing the priorities of the threads, putting the threads into sleep are taken.

Network Bandwidth Controller   Each EE instance creates several outgoing channels (Out-Channels) to send and forward packets. Each OutChannel is responsible for a special combination of a sending protocol, such as ANEP/UDP/IP or ANEP/TCP/IP, and a destination address, which are used for packing the data sent on this OutChannel.

Similar to the CPU resource, when an EE instance is created, the network bandwidth resource that can be used by this EE instance is specified. The controller monitors the network bandwidth resource used by an EE instance by counting the number and the length of the packets sent through each OutChannel, and then summing up the values of each OutChannel. When the total number of bytes sent within a time unit reaches the allowed value in the time unit, the threads used by the OutChannels are prohibited from sending packets for the remaining time in this time unit. When adaptations occur, the bandwidth controller is informed of the newly agreed value, and it controls the number of the packets sent out from the OutChannels according to the new value.

Note that the thread serving the OutChannels is registered in the thread pool belonging to the EE instance. In other words, the CPU resource used for sending the packets are charged to the CPU controller. However, in order to avoid the situation that the network bandwidth resource are blocked by the CPU resource, the priority of the thread used by the OutChannels is controlled by the network bandwidth controller, instead of by the CPU controller.

In addition, similar to some other work related to the resource control in the field of ANs, such as RCANE [Men99] and Janos [THL01], in the GRC module, only the outgoing network bandwidth is considered by the bandwidth controller. The packets arriving at an AN node are received through the incoming channels (InChannels), which are created during the initiation of an EE instance. When creating an InChannel, the following parameters must be specified:

- Which arriving packets are to be delivered on this channel. This is described by a protocol specification, an address specification and a demultiplexing key;

- A buffer pool that queues packets waiting to be processed by the channel;

- A function to handle the packets.

That is to say, each InChannel has a function to process the packets, and a buffer pool that queues packets waiting to be processed by the channel. When a packet arrives at an AN node, it will be put in the buffer pool associated with an InChannel after being checked which InChannel it matches in the demultiplexing phase. Hence, through setting the size of the buffer pool, which is controlled by the memory controller, the incoming network bandwidth can be controlled. Moreover, the threads for handling the packets associated with the InChannels are also under the control of the CPU controller.

In general, the network bandwidth controller does not limit the incoming bandwidth usage artificially. However, in case the buffer pools in the InChannels are full, the arriving packets will be dropped.

Memory Controller   The memory controller restrains the memory usage of an EE instance within the agreed value. The memory used by an EE includes the buffers for packet sending and receiving, program code loading, thread stacks as well as objects and arrays created on the heaps. Similar to the CPU and network bandwidth controller, the pre-accounting method is also used. When the memory used by an EE instance exceeds the agreed value, some action are taken by the memory controller, including trying to recycle some memory already used by some objects belonging to this EE instance and will not be used anymore, or terminating the threads that overuse the memory.

In summary, when a new EE instance is created, a global resource controller (GRC) consisting of a CPU, a network bandwidth and a memory controller is allocated to this EE instance, to manage the resource usage of this EE instance. And the GRC is also registered in the RM module. It is the task of the RM module to manage the GRCs associated with each EE instance.

Resource Management (RM)

This module is the center of the resource management subsystem. It connects other modules with the resource status information in the system. The RM module acts as a coordinator between different modules in the RM subsystem: the admission control algorithm in the P&AC module decides the amount of resources allocated to a new EE instance based on the resource status information provided by the RM module. And the result of the admission control is then passed to the GRC through the RM module. In the meantime, the RM module needs also the up to date resource status information of each EE instance from the GRC module. The main tasks performed in the RM module are summarized as follows:

- Maintain various resource status information in the system.

- Manage the resource controllers in the GRC module.

- Provide necessary information to the GRC and P&AC module.

- Provide resource query functions that can be invoked by EE instances or authorized active applications.

The essence for implementing the above tasks is to acquire the system resource status information, including the total and available resources in the node system and the resources used by each EE instance. Generally, three kinds of methods have been used to obtain the diverse resource status information: test, native library function invocation and accounting. The RM module acquires the total amount of resources that can be used by the node system through some tests at the system initialization and by invoking native OS functions. It obtains the up to date resource usage of each EE instance by querying the resource controllers in the GRC module, where both the accounting and the native function invocation are used to track the amount of resources consumed by each EE instance. We will discuss the resource status determination method in detail in section 6.2.

So far, the basic resource query functions provided by the RM module are at three levels, namely the underlying Windows OS level, the AN node system level and the application level. And they concern the three major types of resources. The resource query functions include appsResUsed(), anNodeResUsed(), and osResAvailable(), osResUsed() etc. After invoking these functions, the resource amount of all the three types of resources is returned. For example, when osResUsed() is invoked, the total CPU, network bandwidth and memory resource consumed in the underlying Windows OS are returned; anNodeResUsed() returns the total amount of resources used by the ANwithARM system, including the system cost and the resources consumed by all the active applications running in the ANwithARM; the appsResUsed() returns a list of all the active applications running within the ANwithARM, together with the amount of resources consumed by each application.

In conclusion, based on Jnodeos, which has provided the general functions of a NodeOS, our NodeOS has integrated an explicit resource management subsystem to realize the adaptive admission control mechanism and to control the resource usage of each EE instance. Note that the division of the three modules in the RM subsystem is only functionally. The communications between these modules are through function calls.

6.1.2   Execution Environment (EE)

The EE is in fact a virtual machine, in which an active application executes. In our node architecture, an EE instance, i.e., a virtual machine together with an application running inside it, is the unit for the accounting of the various types of resources. To date several EE architectures have been defined and implemented, as mentioned in section 2.4.1. Among these we have selected ANTS, namely ANTS2.0, as the basis of our EE. Because it is based on Jnodeos, and takes advantage of the services provided by a Java NodeOS.

6.1.2.1   ANTS

ANTS, an Active Node Transfer System, is a Java-based EE for active applications developed at Massachusetts Institute of Technology (MIT). ANTS provides a toolkit for constructing EEs and applications. It has introduced two novel techniques for programming a network node: the dynamic code loading and a hierarchical protocol architecture, which support the installation of new protocols dynamically in a network node.

Hierarchical protocol architecture

ANTS uses protocols and capsules to organize the processing behaviors in an AN node. Figure 6.3 illustrates the capsule composition hierarchy used in ANTS. In this hierarchy,

- A capsule is a general replacement for a packet. It can carry not only data but also a program and the corresponding parameters. Instead of program and parameters, a capsule can also carry a reference (called method identity) to a program that could be already cached in an active node. For ANTS, programs are Java bytecode.

- A code group is a collection of some functional related capsules, whose programs are transferred as a unit using the code distribution system. Capsules of one code group cannot directly access data or invoke services supported by capsules in another code group.

- A protocol is a collection of related code groups. Capsules belonging to the same protocol will typically manipulate shared information within the network. However, interaction between two protocols in a node is disallowed.

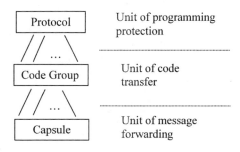

Figure 6.3: Capsule Composition Hierarchy

Hence, using ANTS EE, application developers build a new network service or application by packing their own programs into capsules, combining the related types of capsules to several code groups, which form a protocol. Capsules within a protocol can communicate with each other through states shared in the EE. The EE provides protection by ensuring that capsules belonging to different protocols cannot interfere with each other. In addition, ANTS also adopts a capsule identifier derived from the code description and authenticated by any AN node supporting that protocol to provide security guarantees.

On-demand code propagation

Another technique used by ANTS is the on-demand code distribution mechanism. A code distribution mechanism ensures that a method for processing a capsule can be automatically transferred to the nodes where it is required. ANTS distributes the methods separately from the capsules they are associated with and caches these methods in the network nodes. A lightweight protocol is used to transfer the method code along the path that the capsules follow, in case the method has not already been cached in a node. The advantage is that applications can send capsules into networks at any time, without knowing in advance whether the needed methods for processing them already exist in the network nodes. Figure 6.4 depicts the code distribution procedure in ANTS.

When a capsule arrives at a node, the node checks whether the method needed for processing the capsule is already available in the node. If yes, the corresponding method

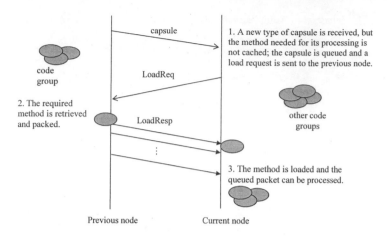

Figure 6.4: On-demand code loading in ANTS

is invoked and the data in the capsule is processed. If not, the capsule is first stored in a queue. Then a method load request message is generated and sent in the LoadReq capsule to the precious node from which the capsule has been received. When the previous node receives the LoadReq capsule, it retrieves the requested method from the local cache and sends it in one or several LoadResp capsules, according to the size of the code and the maximum length limit of one capsule. If the requested method cannot be found, it asks its previous node which is determined according to the information recorded in the LoadReq capsule it has received, namely the source of the capsule who requests the method. Once the current node receives all the LoadResp capsules, the method carried in the capsules is cached in the node. Finally, the required code is available and the capsule in the waiting queue can be processed. If the desired responses are not received in a specified time period, the queued capsules are discarded without further action.

ANTS 1.x are the basic releases that have implemented all the above techniques. They are written in Java and can run in most JVMs directly. ANTSR is a port of ANTS1.1 [WGT98] to Jnodeos by the Janos research group. ANTSR has restructured the internals of ANTS1.1 to take advantage of the services provided by the NodeOS, e.g., to use the primitives like InChannel, OutChannel, CutThroughChannel, Thread, packet filter, Addresses, and Domains defined in the NodeOS API, wherever possible to make the implicit abstractions defined in ANTS more clear. ANTS2.0 is the major upgrade from the last official release from MIT ANTS1.3 and ANTSR co-developed by the University of Washington[2] and the Janos research group. It includes all of the features of both ANTSR and ANTS1.3, plus several new features, including the separation of the dynamic routing

---

[2]The leader is also the researcher of ANTS from MIT.

protocol out of the ANTS core, the introduction of a new security model, which classifies the users into several groups like Admin, Bootstrap, LocalUsers and RemoteUsers and assigns different authorities to the different groups, such as the ability to modify the routing table, read/modify the table of immediate neighbors, shutdown a node, set the node's address and write the log files.

6.1.2.2 Changes to ANTS2.0

ANTS2.0 has taken full advantage of the functions provided by Jnodeos. In order to make ANTS2.0 more suitable to our task, we have made some changes to it.

1. We have added an associated code distribution method to ANTS2.0. The on-demand code loading method limits the distribution of code to where it is needed. However, its startup performance is relative poor. I.e., there exists a relative long processing delay before the "first" capsule can be executed, in case the method for processing the capsule has not yet been cached in the node. The performance of the subsequent capsules that follow the same path and require the same method is much better. This is especially not efficient for capsules requiring only a very short method which can be carried along with the capsule. In order to make our node architecture more flexible and suitable for capsules requiring short programs, we have added an associated code distribution method to the ANTS architecture. Namely, short methods for processing capsules can be carried directly along with the capsules. This method omits the procedure for on-demand code loading. Each capsule carries the method for its processing, and it is particularly suitable for installing and using short programs.

We have defined a new value of Type ID for the ANEP header [ABG+97], representing a new type of EE, to distinguish the two different code distribution methods. User applications may select the code distribution method according to the size of their application code. For the associated code distribution method, each capsule still carries an identifier in the payload field, identifying the method used to process the capsule. This identifier is also used as an index for caching the code in an AN node and processing the capsule. In order to decrease the time used for code loading and caching, when a capsule arrives at a node, the node checks whether the method used for processing this capsule is already available in this node according to the identifier. If yes, the corresponding method is invoked and the data in the capsule is processed, otherwise the code carried in the capsule is first loaded and cached and then the capsule is processed by using the method just cached. The maximum number of cached methods in one node can be defined according to the resources at this node, such as memory space etc.

2. We have made some changes to the format of the capsules. Besides adding a new value of Type ID, we have also defined several new Options in the ANEP header as introduced in the following section. The RV and the ARVS option can be inserted in the packet header, in order to carry the resource requirements and the resource adaptation capability information of an application.

3.   We have added several functions (i.e., interfaces) to the ANTS EE, so that active applications from the end-users can use the resource query functions provided by the NodeOS.

4.   An EE manager has been integrated in the ANTS2.0, whose function is mainly to receive requests from the NodeOS and to decide whether a new EE instance should be created.

5.   Moreover, we have implemented a graphical interface to be used for the configuration of the node system. Therefore, some interfaces and functions have been integrated in the EE to read parameters from the interface, and pass parameters to the NodeOS.

6.   We have also implemented a simple System and Network Management module in ANTS2.0. Its main task is to configure the admission control algorithm and the policy database needed by the P&AC as explained in section 6.1.1.

6.1.2.3   Packet Format

The general format of the active packets traversing our AN node systems is illustrated in Figure 6.5. Basically, we have added a new value to the Type ID field and several options in the ANEP header. The gray fields in this figure emphasize the components that have been added or changed according to our node architecture.

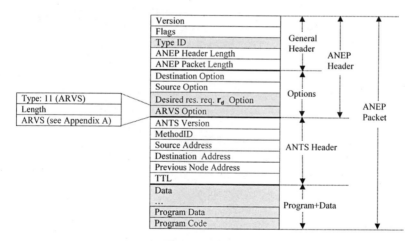

Figure 6.5: Key components of the active packets

The desired resource requirement r_d, which is a RV, and the ARVS parameter of an application are inserted into the ANEP header. Therefore, at each node, just by processing the header of a packet, the P&AC module can obtain enough information for the adaptive admission control algorithm to decide whether the corresponding application can be accepted or not. Both the RV and the ARVS are defined as options [ABG+97] with the structure of TLV (Type/Length/Value) in the ANEP header. So far, we use five types of options. The Destination and the Source option have already been defined in the ANEP. We have defined the RV, ARVS and ConcernedNode (see section 7.1.5) option, using the type identifier 11, 12, 13 respectively. The detail information of these options is given in appendix A. The whole ANEP packet is then stored into a UDP/IP or TCP/IP datagram which traverses networks using the normal IP routing.

6.1.2.4 System and Network Management Module

The System and Network Management module is a mini module used to configure the node system, e.g., to specify the supported EEs, the node address etc. It is implemented as an EE due to unification reasons, but its functions are different from other normal EEs. For instance, it can also change the status and algorithms in the various modules in the NodeOS part. Therefore, it could be considered to be on the border between the EE and the NodeOS.

One important task of this module is to specify whether and how the adaptive resource admission control procedure should be carried out. To be flexible, the admission control procedure can be masked, e.g., in case the active node has been started recently or updated newly and has plenty of resources. Or it can also be configured to operate at a certain level, e.g., at the protocol EE instance level or at the application EE instance level, since the resources in the node system are organized hierarchically, and the ANTS EE, together with Jnodeos, supports the creation of protocol and application level domains for processing packet flows depending on the requirements of the applications. E.g., when simple tasks like forwarding packets are performed, usually only protocol level domains are created, no application level domains are needed; whereas when special or complicated programs such as MPEG encoder or decoder, encryption or decryption are specified by applications, usually application level domains are created to process the different packet flows. So far, the admission control procedure in the ANwithARM can be configured in the following three ways:

- No admission control. In this case, no admission control is performed and no limits for the controllers in the GRC are set.

- With admission control. In this case, the admission control can further be set in two levels: protocol EE instance level and application EE instance level. In the former case, the protocol level EE instances are the unit for admission control, resource accounting and scheduling. In the latter case, we do not care how many resources a protocol level EE instance has consumed. The application level EE instances are the unit for resource accounting, scheduling and admission control.

Other functions of the System and Network Management module include configuring the policy database needed by the P&AC, and registering the needed information in the node system during the initialization of the system. The System and Network Management EE itself is created statically at system start time.

6.1.3 Active Application (AA)

As the application layer in the node architecture illustrated in Figure 6.1, we have implemented a graphical resource monitoring tool and user-node interface, which can be used to monitor the resource usage in the node system and to configure the node system, respectively.

6.1.3.1 Resource Monitoring

To observe the resource usage in the node system, we have implemented a resource monitoring tool in the form of an embedded active application in the node system. A detailed description about this tool will be given in the next chapter, here we address only those issues closely related to the NodeOS and EE.

The tool is implemented as a pure active application. That means, when it is started, an EE instance is created for it, also the number of EE instances in the system increases, and a GRC is allocated to this EE instance and registered in the RM module to schedule its execution. It can use all the services provided by the underlying EE and NodeOS. Particularly, resources consumed by this special application are also counted by the GRC and reported to the RM module. However, the resource limits of this application are set to "infinite". Threads of this application have the same priority as other system threads. Moreover, this application is not subject to the adaptation. The resources consumed by it are charged to the node system. It is also not treated as a candidate application, when adaptations among applications occur. Currently, no r_d and ARVS are set for this specific application. Furthermore, the Protocol ID and Method ID of the application have been registered permanently in the system; therefore, it can directly invoke the resource query functions provided by the NodeOS instead of using the interfaces provided by the EE.

The principle of this tool is simple: the monitoring application polls periodically various resource states in the node system by invoking the corresponding resource query functions directly from the NodeOS. The acquired data is then prepared for graphical presentation and displayed. Therefore, it functionally consists of two parts: data query and display. We will discuss it in detail in section 7.1.4.

6.1.3.2 User-Node Interface

The other tool of the AA part of the node architecture is a graphical user-node interface. Unlike the resource monitoring tool, the user-node interface is not an active application. When it is started, no EE instance is created for it. It is used to configure and start a node system and other active applications. The functions of the user-node interface can be classified into the following two categories:

1. Configuring and starting the node system.

2. Configuring and starting active applications on the node.

In ANTS, the necessary parameters for starting an active node, such as the logical address of this node, the UDP port number for sending and receiving capsules as well as the routing table used by the node etc., are read from a configuration file. To make the node operating more easily and intuitively, we provide a graphical user-node interface so that users can configure the node conveniently. Figure 6.6 illustrates an example for configuring an active node.

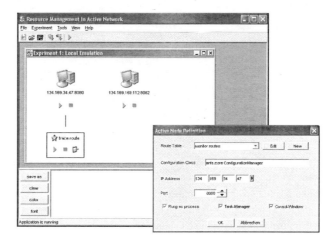

Figure 6.6: Configuring an active node

The parameter settings are then used by the System and Network Management module during the system initialization. In addition, for the purpose of tests and emulations, users can also define several AN nodes in the same machine simultaneously. Each node has its own logical IP address and UDP or TCP port. Different nodes can be started separately, each runs as a windows process.

Furthermore, during the node configuration, users can also specify whether the embedded resource monitoring application needs to be started, as well as whether the prompt information during the system initialization needs to be printed.

In case an active application should originate in this node, the corresponding parameters, such as the desired resource requirement r_d and the ARVS, as well as other application dependent parameters, such as destination address etc., can be given through the interface. The parameters are input using the wizard dialog windows. Figure 6.7 illustrates an example for configuring the r_d and the ARVS.

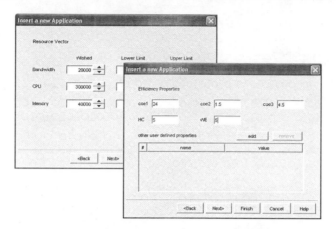

Figure 6.7: Configuring applications

6.1.3.3   Developing Active Applications on the Node Architecture

Since our EE is based on ANTS2.0, methods for developing active applications on top of ANTS [WGT98] are still suitable for our node architecture. The biggest difference introduced by our system is that users can take advantage of the resource adaptation function provided by our node system through specifying the desired resource requirement r_d and the ARVS in the header of the capsule, as shown in Figure 6.5. In addition, application-specified programs, namely the Java bytecode, can also be directly packed in and carried by the capsules throughout the network.

6.1.4   Summary

We have implemented an AN node architecture which supports the suggested adaptive admission control mechanism. The node architecture has used Jnodeos and ANTS2.0 as the basis of the NodeOS and the EE respectively. A RM subsystem is implemented in the NodeOS to realize the adaptive resource admission control through resource accounting and scheduling. In addition, active applications developed according to the principle of ANTS can take advantage of the resource adaptation mechanism in our system by specifying their resource requirements and adaptation capability in the capsules sent to the network.

In this section, we have introduced the RM subsystem in the NodeOS from the perspective of the constructing modules. In the following section, we will present the resource accounting and control schemes used in the resource management subsystem.

6.2   Resource Accounting and Control Scheme

In order to realize the adaptive resource admission control mechanism, the RM subsystem must be able to hold information about the total and available resources in the node system and maintain and control the resources used by each application. The total system resource information depends on the underlying operating system and can be acquired using some native functions. Hence, the main task of the RM subsystem is to monitor the resources used by each application and control them to be within the value agreed during the admission control phase.

6.2.1   General Principle

To keep track of the resources used by each application and control them to be within the agreed value, two rules are followed:

1. An EE instance is the basic unit for resource allocation and control. That is to say, we allocate a definite amount of the network bandwidth, CPU and memory resource to the EE instance, and restrict the total amount of the resources consumed by all the threads belonging to this EE instance within the agreed value. We do not concern about the resources consumed by a single thread of an EE instance.

2. In order to avoid resource overuse, we adopt the pre-accounting technique. That means, measures are taken before executions are performed that will lead to exceeding the agreed limit.

According to the first rule, the logical structure for the resource allocation and control in the RM subsystem is summarized in Figure 6.8.

The resource manager (ResManager) holds all the resource information of the AN node by maintaining variables related to the execution of each application in the system, including the CommonARVS table which stores the adaptability information of applications running in the system. Whenever a new EE instance is created, it is registered in the ResManager and an entry, called Global Resource Controller (GRC) is allocated for it. The agreed resource limit obtained from the admission control will then be set to this GRC. A GRC is responsible for supervising the execution of the threads belonging to the corresponding EE instance. Whenever information related to the system resource status is needed, the ResManager checks all the GRC entries to acquire the up to date states of each application and then returns the needed information after a corresponding calculation.

The GRC maintains the various resource limits of an EE instance and the amount of the resources already used by the EE instance. A GRC consists of a bandwidth controller, a memory controller and a CPU controller. These controllers are in fact system threads. They count the bandwidth, memory and CPU resource used by the corresponding EE

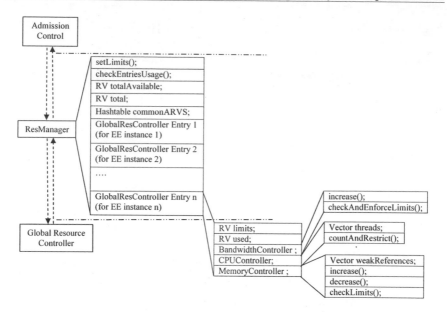

Figure 6.8: Resource manager in RM module

instance and at the same time restrict them to be within the agreed limit value. The detailed information about each controller will be given in the following section.

Figure 6.9 illustrates the procedure of packet processing and summarizes the resources counted to an EE instance on behalf of an application by the resource controllers in the GRC. When an EE instance is created due to the arrival of an application, several types of InChannels are created. Each InChannel is assigned a function to process a certain type of packets, e.g., packets with IP/UDP/ANEP or IP/TCP/ANEP header etc, and a buffer pool that queues packets arriving at this InChannel. The execution of the application program may need data processed by the functions associated with an InChannel; and packets received through one InChannel may need also to be passed to the application program for further processing. As a result of packet receiving, processing and the execution of an application program, packets may be sent through multiple OutChannels, each of which attaches a certain type of transport layer protocol and a destination address to the packets passing through them. Hence, resources involved in the above processing are consumed for the application and counted to the EE instance. I.e., all the related threads are registered in the ThreadPool for this EE instance, and the memory and network bandwidth resource used by the buffers and objects as well as for transmitting packets are counted. Note that the resources for multiplexing/demultiplexing packets to/coming from network devices are shared among multiple EE instances, therefore they are not counted to any EE instance.

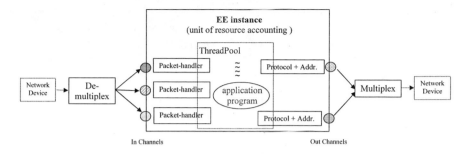

Figure 6.9: EE instance and InChannels, OutChannels

In addition, the ResManager is also responsible for providing the needed information to the admission control. It checks the total available resources in the system whenever there is a request from the admission control module and provides information for the first step adaptation. In the mean time, it checks also the resource usage of all the GRC entries to provide the actual candidate applications for the second step adaptation (i.e., adaptation among different EE instance belonging to the same type of EE instance). The new agreed value after adaptations will be passed back to the ResManager and be set to each GRC through the ResManager, to which the resource controllers conform during the execution of the EE instances. The resource query functions provided by the RM subsystem are based also on these checks.

Concerning the second rule, i.e., pre-accounting resources used by each EE instance, we adopt different methods to count the resources depending on the types of the resources. For memory, we count it before an object, an array or a buffer is allocated, so that the overuse of memory can be avoided. For the CPU and network bandwidth, we divide the restriction interval[3]into small time slices, and count the usage in each time slice. If the resource usage in one time slice exceeds the assigned value, actions such as lowering the priorities of the threads, suspending or evening stopping the threads, will be taken, so that the total CPU and bandwidth usage in the restriction interval will not exceed the agreed value.   .

6.2.2   Resource Determination Method

In principle, two methods are used by the GRC to acquire the resource usage information: native library calls and program variables interception. Through native functions, as introduced in section 6.2.3, the total amount of system resources can be achieved; and by inserting a piece of code in some key points, the needed resource usage information can be acquired.

---

[3]In our system, the CPU resource is specified as bytecodes per second, and bandwidth is bytes per second, therefore the restriction interval is one second.

Generally, the code for accounting is inserted at two levels: source code level and Java bytecode level. Resources used by system threads and functions on behalf of an EE instance, such as NodeOS threads created for each EE instance for packet reception, exception processing etc., can be acquired at the source code level, since we can manipulate the source code of this part. In addition, some resources, such as the network bandwidth, can only be used by an application through invoking the system provided functions, e.g., OutChannel.sent(). Therefore, this part of resources can also be counted at the source code level.

The CPU and memory resource consumed by threads for executing application programs[4] have to be counted at the Java bytecode level. On the one hand, it is unrealistic to expect the source code of each application to be available for modifications. E.g., in our node system, the code injected by active applications is Java bytecode; we cannot count the resource usage at the source code level. On the other hand, we cannot rely on the application programmers to notify us about the resource usage of their programs. Hence, we adopt the Java bytecode engineering technique to insert pieces of code to the application programs to count the resource used by the application programs.

The rational of bytecode engineering is to allow the manipulation of already compiled Java programs, and modify directly the Java bytecode without resorting to any external source compiler or source code. Bytecode rewriting can change the compiled Java ".class" files directly, insert pieces of code into the original bytecode before it begins to execute. In addition to changing the bytecode in advance (off-line rewriting), the bytecode can also be rewritten on-line. On-line rewriting occurs after the JVM loads the class file and before defining the class. After rewriting is finished, the new byte array is passed to the function defineClass(), and the new class is returned to the JVM, just as if it were the original class. Figure 6.10 illustrates this procedure.

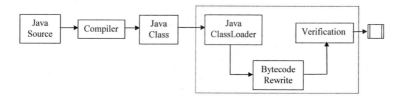

Figure 6.10: On-line bytecode rewriting

There are several bytecode engineering frameworks written in Java, such as BCA [KH98], BIT [LZ97] and Javassist [Chiba00]. We have chosen BCEL (Byte Code Engineering Library) [Dahm99] [BCEL] for the bytecode rewriting to insert the resource accounting information. BCEL allows fine-grain manipulation of Java bytecode. It allows arbitrary

---

[4]Here the programs include also those methods which has already cached in the node system when the on-demand code distribution method is used.

modification of the class files, even the construction of new classes from scratch at byte-code level. BCEL consists of a static and a generic part. The static part is not intended for bytecode modification and can be used to analyze Java classes directly from the class file. The generic part supplies an abstraction level for the creation and transformation of class files in a dynamic way.

Figure 6.11 illustrates the general procedure for bytecode rewriting by using the BCEL framework. First the class needed to be rewritten is retrieved from the repository and a JavaClass object is formed which contains all the information about the original class. Based on the JavaClass, the class, constant pool and methods as well as the bytecode instructions of each method can be obtained. Therefore, the position where the code for accounting should be inserted can be easily found, e.g., just before the allocation of an object (where the bytecode instruction "new" is used and the constructor of an object is invoked), and the corresponding code for accounting can be inserted. Finally, the modified class can be generated.

```
JavaClass jc=Repository.lookupClass(classname);
ClassGen cl=new ClassGen(jc);
ConstantPoolGen cp=cl.getConstantPool();
  //extract methods from the class
Method[] mths=cl.getMethods();
  // insert code to methods one by one
for (int i=0; i<ms.length;i++) {
    MethodGen mth= new MethodGen(mths[i], classname,cp);
    modifySignature(mth);
    //extract the instruction list from the method
    InstructionList ins=mth.getInstructionList();
    //find the position to insert, e.g., before objects are allocated
    InstructionList newIns=ins+addCount;
    //update the method using the new instructions
    mth.getInstructionList(newIns);
    //replace the method with the new one
    cl.replace(mths[i], mth.getMethod());
    cp.addMethodref(mth);
}
  //generate the rewritten bytecode
jc=cl.getJavaClass();
jc.setConstantPool(cp.getFinalConstantPool());
```

Figure 6.11: General procedure for bytecode rewriting

In the following, we discuss how the bandwidth, memory and CPU controller in the GRC work to count the different kinds of resources consumed by the active applications and simultaneously restrict them to be within the agreed limits. In order to improve readability, we explain our methods at the source code level, even through some of them are implemented through bytecode rewriting.

6.2.3   Resource Accounting and Control

Due to the unique features of each type of resource, we have to count and control the usage of each type of resource separately. In the following, we analyze the usage of each type of resource in our node system and present the methods for controlling them in each resource controller. Basically, each resource controller has two main functions: one is to count the resources used by each EE instance, and the other is to restrict the resources within the agreed value.

6.2.3.1   Network Bandwidth Controller

As mentioned above, packets arrived at an AN node are received and processed by InChannels. Each InChannel created for an EE instance has a buffer pool that queues packets to be processed by the InChannel. Hence, through setting the size of the buffer pool, the incoming network bandwidth can be controlled. Therefore, the bandwidth controller has an impact on the outgoing network bandwidth only.

An EE instance sends packets using one or several OutChannels. Each OutChannel specifies a destination address and a transmission layer protocol, which define the headers attached to each packet sent through this channel. E.g., if UDP is specified in an OutChannel, the ANEP packets sent through this channel are first packed into UDP and IP datagrams using the destination address specified in this OutChannel and then sent out. Hence, the network bandwidth consumed by an EE instance consists of the bandwidth used by all the OutChannels created by this EE instance. Moreover, packets sent through the OutChannels may include the packets sent from the source application and processed and forwarded by the node as well as packets possibly generated by the code execution in the node.

Accounting of the outgoing network bandwidth is relative straightforward. The sum of the number of the packets sent through each OutChannel belonging to the EE instance and the length of the packets can be directly recorded by the bandwidth controller, and both the system and application threads can only send packets through these OutChannels. The unit of network bandwidth is bytes per second. Note that here the packet length means the total length of the UDP datagram or TCP segment, including the header of UDP or TCP.

In order to prevent the bandwidth usage from exceeding the limit, we divide the 1 second interval into several slices and restrict the number of packets sent in each time slice according to the result of admission control[5]. The smaller the time slice is divided, the more precise the bandwidth usage can be controlled, but the more overhead is introduced. Hence, balancing the overhead and the result of resource control, we set the time slice to 100ms for the moment. Namely the amount of packets sent out by all the OutChannels belonging to one EE instance is counted and actions are taken according to the 100ms

---

[5]The permitted value in one time slice is the result from admission control divided by the number of time slices in one second.

criterion. Figure 6.12 illustrates the main variables in the bandwidth controller and the piece of code for counting and restricting the bandwidth usage inserted right before the packet is sent. We use two values in the bandwidth counter to record the data in bytes that an EE instance sent. One is totalSent, which records the total data that the EE instance sent. The other is intervalSent, which records the data sent in a certain time (may be shorter or longer than 100 ms). When a packet is sent, both values increase. Each time when intervalSent exceeds the permitted value, the controller checks how long it has elapsed. If it is shorter than 100 ms, which means more bandwidth will be used in the 100ms time slice than the agreed value, the threads for sending packets are put into sleep until the next 100ms-interval. Then intervalSent is reset to 0 and the following packets sent are counted to next 100ms-interval. If the elapsed time is longer than 100ms, intervalSent is reset to 0 and the packets sent are counted to the next 100ms-interval.

```
public class bandwidthController {
    long totalSent;  //total number sent
    long intervalSent;  //bytes which may be sent in one 100ms time interval
    long intervalBeginTime;  //begin time of one 100ms time interval
}
//following piece is inserted where packets are sent
...
sentBytes=packetLength;  //number of bytes that will be sent
bandwidthCount.totalSent+=sentBytes;
bandwidthCounter.intervalSent+=sentBytes;
if (bandwidthCounter.intervalSent >=limit ){
    currentTime=getCurrentTime();
    elapsedTime=currentTime-intervalBeginTime;
    if (elapsedTime<100ms ) {
        sleepTime=100 –elapsedTime;
        sleep (sleepTime);
    }
    intervalSent=0;
    intervalBeginTime=currentTime;
}
send(sentBytes);  // then sent the packet
```

Figure 6.12: Pre-accounting network bandwidth

As shown in Figure 6.12, the pre-accounting principle is followed. I.e., first the value in the counter is increased and the limit is checked. If the limits will be broken, measures are taken to prevent the resource overuse. Through this way, the network bandwidth actually used by an EE instance cannot go beyond the agreed value.

Note that we put the threads for sending packets into sleep whenever the bandwidth usage will exceed the limit. This method avoids discarding packets, and the CPU resource for this EE instance may decrease, however, the memory usage may increase for buffering the packets to be sent.

6.2.3.2   CPU Controller

The CPU resource needed by an application in the node system is spent by threads for sending and receiving packets, as well as executing the application-specified program. Threads for packet receiving and sending are initiated by the NodeOS for InChannels and OutChannels, as the application is accepted by the node system and an EE instance is created for it. The application program may initiate multiple threads during the execution, each thread performs one part of its task. During the execution, system functions can be invoked. Since all these threads run on behalf of an application, the CPU cycles used by these threads are counted to the corresponding EE instance.

The unit for applications to specify their CPU requirement and to control the CPU usage in ANwithARM is bytecodes per second (BCPS). The advantage of using BCPS is that given the BCPS requirement, the execution time of an application can be predicted independent of platforms. This is particularly important, because the active applications may always travel through heterogeneous systems, i.e., based on different hardware and providing different performance.

However, using BCPS requires theoretically the CPU usage of an EE instance to be also counted in bytecodes. In other words, the number of bytecode instructions executed by all the threads running on behalf of an application have to be counted. This can only be implemented through bytecode rewriting technique. I.e., code for counting the number of bytecode instructions executed by the threads must be inserted into both the system and the application program at bytecode level. Because the system program is complex and the specific characteristics of the various application programs cannot be anticipated, e.g., there may be many branches in the program (e.g., caused by the if and switch statements), the accounting code must be inserted before each block of the bytecode instructions between two branches. This increases greatly the overhead for CPU resource accounting. Hence, we adopt native OS methods to obtain the CPU time used by the threads to calculate the CPU usage of each EE instance instead of counting the number of the executed bytecode instructions directly. However, to keep the platform independence, the following key problems have to be solved:

1. How to convert the amount of CPU usage expressed in BCPS to that expressed in CPU time and vice versa?

2. How to map the Java threads in the ANwithARM to the threads of the underlying operating system (Windows OS threads), so that the CPU time used by the Java threads can be obtained?

3. How to control the CPU usage with limits specified using BCPS?

To solve the first problem, we use the system performance expressed in BCPS. The node system performance expressed in BCPS can be calculated through measuring the elapsed time after the execution of a known number of bytecode instructions during the system initialization. According to the system BCPS, the CPU time consumed by an EE instance can be converted to the number of the executed bytecode instructions and vice versa. Of course, since the system BCPS is acquired through measuring the elapsed time of executing a mixture of bytecodes, there may be some errors during the conversion. To remedy the errors, we allow for some relaxation of restrictions during the CPU usage control.

The second problem is solved by invoking a native method. Within the NodeOS, it can be examined whenever a thread belonging to an EE instance is created. Therefore, after the initiation of a thread, a native method is invoked to record the thread identifier (ID) of this thread in the underlying operating system. Thus, an operating system level handle is created for this new Java thread. And this handle is registered in the CPU controller, which queries the underlying OS periodically about the CPU time used by this thread. Hence, the total CPU usage of an EE instance can be acquired through summing the CPU time used by all the threads registered in this CPU controller.

The third problem is also one of the main tasks of the CPU controller. To control the CPU usage of an EE instance, we use a high-priority thread in the CPU controller which queries periodically the CPU time used by the EE instance and calculates the number of bytecode instructions executed using the system BCPS. In order to limit the total bytecode numbers within the agreed amount, we set a 10ms time slice. The counter checks the bytecode instruction number executed by each EE instance every 10ms. If the number exceeds the agreed value, actions such as lowering the priorities of the threads or suspending the threads are taken.

Figure 6.13 illustrates the algorithm for restricting the CPU usage of each EE instance in the CPU controller. The CPU controller first calculates how many bytecode instructions an EE instance is permitted to execute in a 10ms time slice according to the result of the admission control expressed in BCPS. Then it invokes the method countAndRestrict() every 10ms to enforce that the number of bytecode instructions executed by all the threads belonging to the EE instance in a 10ms time period does not exceed the agreed value. If the number exceeds the agreed value, the counter lowers the priority of all the threads belonging to this EE instance. In the meantime, depending on the degree that the agreed value is exceeded, the threads of the EE instance may also be put into sleep for one 10ms time slice. If the value is below the agreed value for some degree (e.g., now 20%), the priority of the threads is increased. However, the priority will not exceed the original one allocated to the EE instance as it has been created.

To a certain degree, the use of time slices for CPU controlling can be considered as a method for pre-accounting. Because we have not adopted the bytecode rewriting technique for CPU accounting, we cannot increase the counter just before a block of bytecode instructions is executed and check if the limit will be broken. Note that here the 10ms time slice is not fixed; it is chosen considering the balance of the overhead introduced by the CPU accounting, the precision of the CPU time for each thread acquired through the native methods (now the precision is 100 nanosecond), and the principle of

```
Vector threadsInfo;  // stored information about all the threads related to one EE instance
long sysBCPS;  //system's average bytecode number per second
long total;  //total bytecode number executed in the current time slice
long lastCountTotal;  // total number executed in the last counting time slice

void countAndRestrict() {
    while ( hasMoreElements() ) {
        thread = threadsInfo.nextElement().threadHandle;  //get the handle of a thread in the vector
        totalTime=getThreadTime(thread.threadOSID);
            //native method, get the time the thread has used since it is created.
    }
    total=totalTime*sysBCPS;
    if (total-lastCountTotal>2*agreedLimit) {
        threadsInfo.setSleepNotify(10^5);   //10ms
            //set a sleep sign to all the threads with the indicated time (the duration of a time slice)
        threadsInfo.setPriority(getPriority()-1); //lower the priority of all the threads
    }
    else if (total-lastCountTotal>1.2*agreedLimit) { //20% higher than the agreed value
        threadsInfo.setPriority(getPriority()-1); //lower the priority of all the threads
    }
    else if (total-lastCountTotal<0.8*agreedLimit) { //20% lower than agreed value
        if (getPriority()<(original-2) )  // maximal allowed 2 higher than the original one
            threadsInfo.setPriority(getPriority()+1); //increase the priority of all the threads
    }
    // when the executed value is within (1±20%)*agreed value, do nothing
    lastCountTotal=total;
    totalTime=0;
}
```

Figure 6.13: CPU Controller

pre-accounting.

In addition, threads used by OutChannel for sending packets are counted to the CPU controller, however, their priority are controlled by the bandwidth controller instead of the CPU controller. Through this method, on the one side the accuracy of CPU accounting is not decreased, on the other side the bandwidth usage is not affected. The disadvantage is that the CPU resource cannot be controlled in the case that the allocated CPU resource for an EE instance is small but the bandwidth resource is large. So far we have not considered this special case.

6.2.3.3 Memory Controller

During loading the application-specific code, receiving, processing as well as sending packets, memory is needed for packet buffers, storing code modules, thread stacks, objects and arrays. In our node system, packet buffers have a fixed length based on Jnodeos, and memory for code modules can be counted according to their length. The amount for thread stacks are the product of the number of the threads and the maximum stack size[6] set when a thread is created in the system. The variation of these values is

---

[6]Our underlying JVM allocates execution stacks that cannot expand dynamically. In case the JVM allows execution stacks to enlarge significantly, stack size for each method invocation should be counted, which is

recorded in the memory counter by increasing or decreasing the totalUsed field in the memory controller (see Figure 6.14). Objects and arrays created by the execution of the application programs and processing packets use heaps. Due to the garbage collection mechanism of the Java Virtual Machine (JVM), accounting of the heap memory used by an EE instance is a little complicated.

Garbage collection is a kind of memory recycling technique used by the JVM. When an object is no longer referenced by a program, the heap space it occupies is recycled and made available by the underlying JVM for subsequent new objects. Therefore, during accounting the memory used by an EE instance, we have to consider the amount once allocated to the EE instance, but recycled by the system and which may be again allocated to the EE instance.

Hence, we use a method similar to [CE98] to count the heap memory used by the objects and arrays of an EE instance. Namely, for the non-array objects, the constructors are modified to register the reference of the memory controller that the objects belonging to and the finalizers are either modified or generated so that the memory controller can be notified when an object is freed by the garbage collector. In the mean time, a simple method is inserted just before the object allocation instruction. The inserted method increases the counter by the object size, and verifies that the memory limit will not be exceeded. The object size is determined by the fields of the Java basic types, the fields holding object references, plus a constant for the object overhead. For the arrays, after each array allocation in the original method, a weak pointer for this array object is registered in a table of the memory controller. This is because the weak pointers do not obstruct de-allocations of objects. When an array object is not used, and garbage-collected, the corresponding weak pointers will become null. Thus, when the memory counter detects the null pointer, the corresponding size can be subtracted from the usage field. Note that in principle, non-array objects can also use the technique of weak pointers, however, this increases the overhead for maintaining the weak pointers.

The pre-accounting rule is adopted for the memory accounting. As shown in Figure 6.14, the memory controller maintains a limit checking method (checkLimit()), which is invoked just before the memory is allocated to an EE instance. Suppose the amount newSize should be allocated to an object, then the method first checks in advance if the limit will be broken when newSize is allocated to the EE instance. If yes, the method checks whether some memory has already been or can be recycled. It first examines if there exist objects that have already been garbage-collected but their memory amount has not yet been subtracted from the counter. If yes, the corresponding amount of memory is subtracted and the limit is checked again. If the limit is still broken, the method initiates the garbage collection procedure to try to recycle some memory and see if there is recycled memory ever used and can be reuse by this EE instance. If the total usage still exceeds the limit, the limit checking declares fail. In our current implementation, in case the limit

---

decided by the local variables of the method and the size of each frame created for the method invocation.

check is not passed before any memory allocation, the thread allocating the memory is terminated.

```
public memoryController {
    public long totalUsed;
    public Hashtable occupyRefSet=new Hashtable(); //to save objects that are managed by the counter
    public final static ReferenceQueue countRefQ = new ReferenceQueue();
            // to save info if an object has been garbage collected

    public boolean checkLimit ( int newSize) {
        if (totalUsed+ newSize >limit) { // in case memory with the length newSize is allocated, if limit will be broken
            while ( ref = (CountRef) countRefQ.poll() !=null) {
                    //check if there are objects having been garbage collected
                if (occupyRefSet.remove(ref))  //check if ref still recorded in the reference set
                    totalUsed-=ref.size;   //the object has been garbage collected, therefore the used
                                           //memory could be reduced from the counter
            }
            if ( totalUsed+ newSize >limit) { // still break the limit
                System.gc();   //initiated actively garbage collection
                while ( ref = (CountRef) countRefQ.poll() !=null) { //check if objects have been newly gc
                    if (occupyRefSet.remove(ref))
                        totalUsed-=ref.size;
                }
            }
            if (totalUsed+newSize>limit)   // after initiating gc, still break the limit. Limit check fail
                return false;
        }
        totalUsed+=newSize;
        return true;
    }
}
```

Figure 6.14: Memory controller

Note that the garbage collection costs also CPU cycles. Since the single heap system is used in our AN node system, the garbage collector recycles memory for all the EE instances and the system, we cannot calculate how much time is spent for one EE instance. Hence, so far we have not assigned the CPU cycle cost used for the garbage collection to an EE instance. To reduce the CPU cost, only when the excess of the memory usage is detected by the memory controller, the garbage collection is started. However, this may also introduce some errors, namely it is possible that some objects are garbage-collected by the system, but this cannot be reflected on time in the counter of the memory controller. We will discuss this case in the following section.

6.2.4 Summary

Section 6.2 has introduced the resource management scheme used in the ANwithARM system. Each EE instance is allocated a global resource controller to monitor the bandwidth, CPU and memory resource used by the EE instance, and control them to be within the value agreed during the admission control. Both the native library invocation and bytecode rewriting method are used to count the resource usage, and the pre-accounting technique is adopted by the resource controllers to prevent the resource usage from exceeding the limits. In the following section, we will evaluate the performance of the resource controllers and show that the overhead and errors introduced are acceptable.

Nevertheless, several factors have not been considered yet in the ANwithARM system, e.g., the co-accounting and co-controlling of the different types of resources, and the CPU cost for garbage collection.

6.3   Evaluation

In the following, we evaluate the above resource accounting and control scheme from two perspectives: the accuracy of the resource accounting, and the overhead introduced by the resource controllers. In addition, we examine also the associated code distribution method introduced in the EE by comparing it with the on-demand code distribution. The evaluation of the whole ANwithARM node system from the point of view of the active applications is given in chapter 7.

6.3.1   Accuracy

In ANwithARM, the bandwidth resource used by each active application, i.e., the number of packets and their length sent by each EE instance can be counted well with the help of Jnodeos. Hence, in the following, we evaluate only the accuracy of the CPU and memory accounting in GRC.

6.3.1.1   CPU Accounting

As mentioned above, using the system BCPS and the CPU time consumed by the threads of an EE instance to estimate the number of bytecode instructions executed by these threads may introduce some errors. In order to observe the errors introduced, we have performed tests to measure the difference between the estimated and the actual number of bytecode instructions executed. We inserted pieces of code into the test program to count the number of bytecode instructions actually executed by the threads of the EE instance representing the application through the bytecode rewriting technique, and simultaneously record the number of bytecodes inserted for accounting to measure the overhead introduced. This has then be compared with the result obtained by using the estimation method based on the system BCPS. We have used the Image application developed on top of the ANwithARM (details about this application is given in section 7.1) as the test application.

We have made the tests using computers with Intel Pentium(R) III mobile CPU running at 650MHz and 866 MHz and Pentium(R) IV CPU running at 2.0GHz respectively. First, we adopt the estimation method. We let the Image application run in each computer separately and record the CPU time used by the related EE instance for processing the image request capsule and sending all the image data to one who has requested it, so that the executed bytecode instruction number can be calculated by multiplying the CPU time and the system BCPS. Then, we replace the original Java program of the Image application with the rewritten one. Namely, we inserted in advance pieces of code for counting the number of the executed bytecode instructions into the original Java program offline using the bytecode engineering technique, and then run them in the ANwithARM node system. Thus, during the execution of the rewritten Java program, the number of the bytecode instructions executed is counted and recorded. Figure 6.15 illustrates the results of these tests in each computer with the maximum, middle[7] and minimum error. Note that here the counted number of bytecode instructions executed does not include the number of the inserted bytecodes for accounting, they have been subtracted. To observe the overhead introduced by counting the number of bytecode instructions directly using bytecode rewriting method, we have also recorded the executed bytecode instructions inserted for accounting, which is shown in the last column of Figure 6.15 in absolute numbers and in percent of the total executed bytecode numbers.

Computer performance	Test	CPU time used (ms)	Bytecode No. estimated	Bytecode No. counted	Error (%)	Bytecode inserted	
						No.	inserted/counted(%)
Pentium III mobile 650MHz sysBCPS=25402454	min.	459.0080	11659930	11303264	3.2	926862	8.2
	mid.	461.7920	11730650	11303264	3.8	926862	8.2
	max.	471.5424	11978334	11303264	6.0	926862	8.2
Pentium III mobile 866 MHz sysBCPS=30057737	min.	360.8064	10845023	11303264	-4.1	926862	8.2
	mid.	372.2348	11188536	11303264	-1.0	926862	8.2
	max.	381.0144	11452431	11303264	1.3	926862	8.2
Pentium IV 2.0GHz sysBCPS=72655263	min.	157.6624	11455003	11303264	1.3	926862	8.2
	mid.	161.8208	11757132	11303264	4.0	926862	8.2
	max.	165.2448	12005904	11303264	6.2	926862	8.2

Figure 6.15: Number of bytecode instructions estimated and counted

From Figure 6.15 we can see that there are some difference between the estimated and the counted value (from -4.1% to 6.2%). This is because on the one hand, the CPU time counted by the CPU controller may be a little different due to some interior factors, such as the state of the system when the image request is received. On the other hand, the accuracy of using the system BCPS to estimate the bytecode number executed by an application depends on the "similarity" of the application code and the system code used to measure the system BCPS. The background is that executing different bytecode instructions may consume different CPU time. E.g., the bytecode instruction *iconst_0* only pushs the constant 0 onto the stack; whereas *if_icmpeq* pops two integer values from the stack and then compares them. Both of them are counted to one bytecode instruction. Nevertheless, the CPU cycles they consume are different. Because we could not simulate the bytecode combination of each active application during the measurement of the system BCPS, this error is unavoidable.

---

[7]We have performed 12 tests in each computer, the middle value is the average value of 10 tests except the maximum and minimum one.

However, compared with the complexity and overhead introduced by the bytecode rewriting method, we still selected the estimation method to count the number of bytecode instructions executed. The overhead would be do large because there may be many statements like *if, switch, exception throwing* etc. in the application program, which cause many branches at the bytecode level. And between each two branches, minimal 6 bytecode instructions for accounting, as shown in Figure 6.16, have to be inserted. Depending on the number of the branches in the application programs, the overhead may be different. E.g., for the Image application, the overhead of the inserted code is 8.2%. Note that here we have not measured the CPU overhead for inserting the bytecodes for accounting.

```
aload 1         //load reference to the accounting object counter
dup             //duplicate the value on stack to store the result
getfield  BytecodeCounter::totalNbr
                // load the value of totalNumber field of counter
bipush 12       //put the amount of bytecode instructions to be added, e.g., 12
iadd            //add the value: totalNbr +12 on stack
putfield  BytecodeCounter::totalNbr
                // store the added value back to the counter
```

Figure 6.16: Inserted bytecode instructions for accounting

6.3.1.2   Memory Accounting

To evaluate the errors of the memory accounting method in GRC which are introduced due to the garbage collection, we have designed a test memory counter to count the memory used by the EE instances simultaneously with the memory controller in GRC. The test counter invokes the system garbage collection each time when the counter should be increased due to, e.g., the allocation of new objects (and thus consumes also extra CPU resource). As test application, we have selected the Traceroute application which does not perform any special processing except collecting some simple information about the node system (it will be introduced in detail in the next chapter). Since the Traceroute application allocates objects only to consume the allocated memory without any other use, the objects allocated are garbage-collected by the system very often. Hence, it represents an extremely worse case for testing the memory controller. We compare the value recorded by the counter in the memory controller with that acquired by the test counter to observe the errors of the memory accounting. Figure 6.17 illustrates the memory usage acquired by the memory controller and the test counter during the execution of the Traceroute application in the machine with Intel Pentium(R) III mobile CPU running at 866MHz, denoted as the counted and actually used values respectively. Note that here the Traceroute application has a memory usage limit of 21016 bytes.

From Figure 6.17 we can see that the counter in the memory controller increases

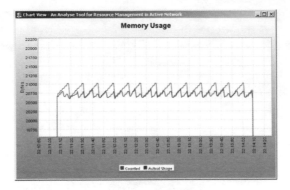

Figure 6.17: Memory value counted and actual used

whenever new memory is allocated. After the memory controller detects that the memory usage will exceed the permitted value, it checks if there are some objects already garbage-collected by the system. If yes, the corresponding amount is subtracted from the counter. This procedure repeats until the application ends. The curve acquired by the test counter varies more heavily than that acquired by the memory controller, because the test counter invokes the system garbage collection more frequently than the memory controller. Due to the garbage collection, there are some differences between the value recorded by the memory counter and the test counter. In this example, the maximal difference is 296 bytes. However, similar to the CPU resource, the difference depends on the execution of the concrete active application, where the life time and the memory size of the objects may vary greatly.

6.3.2   Overhead of Resource Controllers

In order to evaluate the overhead introduced by the bandwidth, memory and CPU controller, which both count and control the usage of the corresponding resources, we have carried out several experiments to measure the execution time of an application in the system with and without these controllers.

We run the simplified Image application in the test active network consisting of our node system. A detailed description of this application will be given in section 7.1; here we explain only those aspects related to the overhead measurement. A user sends a request capsule to a server to ask for an Image, the server makes response to the request by sending the data of the desired image in several response capsules. In the Image application described in section 7.1, the format of the image sent back to the user is determined according to the resource status in the server node system. In the following tests, in order to keep the application execution with and without resource controllers as similar as possible, and to measure the execution time for sending an image with the same size, we have simplified the Image application. Namely in the server, the format of the image sent is not selected dynamically. The image data is sent out in the original PNG format. We measure the interval between the time when a request capsule is received and begin to be processed by the server node system and when all the image data is sent out in the response capsules. In our tests, the server active node is run on a computer with Intel Pentium(R) III mobile CPU running at 866 MHz.

In the first test, we disabled all the resource counters in the node system. In the second test, all the three counters were enabled, but there were no limits on the resource usage. That means the resources used by the application have been counted by each resource controller, and the resource control functions in each counter have checked each time whether the resource limit is broken, nevertheless, no real actions have been taken since no resource overuse has occurred. I.e., in this case, no further overhead is introduced by the resource usage limitation, such as putting threads into sleep etc. Hence, the overhead introduced by the resource counters can be well measured. Figure 6.18 illustrates the results of several tests.

Without Counters		With Counters		Average overhead
Test No.	Time (ms)	Test No.	Time (ms)	
1	1082.8432	1	1102.3488	(average$_{without}$
2	1071.5408	2	1114.7936	- average$_{with}$)/
3	1031.4832	3	1081.3536	average$_{with}$
4	1051.1520	4	1097.8576	
5	1068.4112	5	1082.3824	
average$_{with}$	1061.0861	average$_{without}$	1095.7472	3.3 %

Figure 6.18: Overhead of the resource controllers

From Figure 6.18 we can see that the resource controllers introduce about 3% overhead for the simplified Image application. In practice, the overhead may depend on application programs, e.g, the number of objects allocated, the number of packets sent etc.

6.3.3   On-demand and Associated Code Distribution

As mentioned in section 6.1.2, we have added an associated code distribution method to our EE. Therefore, in ANwithARM, in case the required code for processing a capsule cannot be found, an on-demand code loading procedure is initiated or the required code is loaded from the capsule payload according to the Type ID value specified by the application in the packet header. In order to show the necessity and the performance of the associated code distribution method, we have performed several tests using the Traceroute application. We examine the processing time for each capsule in the network node as well as the round trip delay of the capsule, i.e., the total transmission delay, when different code distribution methods are used.

As it will be introduced in chapter 7, the Traceroute application can record the time point when each TracerouteCapsule arrives at (t_{arrive}) and leaves (t_{leave}) an active node. Thus, the processing time for each capsule at the AN nodes and the total transmission delay can be obtained. Figure 6.19 shows the path of the TracerouteCapsules for these tests. The packets sent from the source node traverse the intermediate node to get the destination, and they come back to the source through the same intermediate node again. I.e., they transit the intermediate node twice. In each test, we send 30 TracerouteCapsules at an interval of 2s at the source node, and calculate the processing time experienced by each capsule using $t_{leave} - t_{arrive}$ when it passes through the intermediate and the destination node. The total transmission delay is calculated according to the interval between the time when a capsule leaves and returns to the source node. In order to reduce the affects of the resource adaptation on the processing time, in these tests we let each node have abundant resource so that the Traceroute application can pass the admission control without any adaptation. Moreover, we eliminate the code for processing the Traceroute application cached in the intermediate and the destination node before each test, such that when the first capsule arrives at the nodes for the first time, no required code can be found and a code loading procedure must be invoked.

Figure 6.19: Capsule path in the test network

Because in practice there is a limit to the maximum length of each packet, the code may be transferred in several segments. In order to examine the relationship between the performance of the code distribution method and the code length, we simulate the cases of different code length by limiting the size of the UDP transmission buffer in the program, such that the same code can be transmitted in different numbers of UDP segments. Clearly here we have neglected the errors caused by the MTU for transmitting packets of different length.

Generally, the processing time needed by a capsule at a network node may include four parts: time used for deciding whether a new EE instance should be and can be (through admission control) created, searching and possibly loading and caching the required method, creating a new EE instance, and processing the capsule using the specified method. For the first capsule traversing a node for the first time, i.e., in the cases when the first capsule arrives at the intermediate node on the way from the source to the destination and arrives at the destination node, a new EE instance needed to be created and the needed code for processing the capsule has not been cached in the node. Therefore, a code loading procedure occurs. When the on-demand code distribution mechanism is adopted, the method will be fetched from the previous node. And when the associated code distribution mechanism is used, the code is obtained directly from the capsule body. For the following capsules and for the first capsule traversing the intermediate node for the second time, i.e., on the way back to the source from the destination, an EE instance for this application has already existed. No admission control and code loading procedure are needed.

	Code Segment Number N (in UDP Layer)	Capsule Number	Intermediate node for 1st time (ms)	Destination node (ms)	Intermediate node for 2nd time (ms)	Total transmission delay (ms)
On-Demand Code Load (code length=4179 bytes)	N=1	1st Capsule	71.662	121.560	1.002	257.336
		2nd Capsule	0.890	3.334	0.922	61.242
		3-30th Capsule	1.004	3.344	1.152	61.528
	N=2	1st Capsule	95.034	144.236	1.112	302.118
		2nd Capsule	1.032	4.434	1.308	63.452
		3-30th Capsule	1.012	3.401	0.952	62.118
	N=5	1st Capsule	184.212	235.340	1.350	480.890
		2nd Capsule	1.118	3.674	1.342	63.238
		3-30th Capsule	1.067	3.680	1.360	63.332
Associated Code Load (code length=6343 bytes)	N=1	1st Capsule	20.668	70.334	2.842	157.112
		2nd Capsule	3.002	6.526	2.868	75.026
		3-30th Capsule	2.806	6.796	2.900	74.696
	N=2	1st Capsule	42.326	100.118	10.328	220.496
		2nd Capsule	8.716	14.662	9.224	98.588
		3-30th Capsule	9.398	14.254	10.136	100.196
	N=5	1st Capsule	133.444	198.228	45.232	457.226
		2nd Capsule	48.118	52.324	46.556	230.190
		3-30th Capsule	48.506	54.072	47.342	232.192

Figure 6.20: Processing time using on-demand and associated code distribution

Figure 6.20 illustrates the results of the processing time and total transmission delay for each capsule when the two code distribution methods are used. The values are an average of 5 tests. The values for the capsule number from 3 to 30 are averages for these capsules of 5 tests. From this figure we can see that when the first capsule arrives at the intermediate node for the first time and arrives at the destination node, in other words, the required code has not been cached in the nodes, the processing time using the on-demand code distribution mechanism is much larger than using the associated mechanism. I.e., the startup performance of the on-demand code distribution mechanism is not as good as that of the associated mechanism. For the following capsules, the experienced processing time and total transmission delay using on-demand mechanism is better than using the associate mechanism. However, the difference is not large when the number of the segments of the program is equal to 1, i.e. all the code can be transferred in one capsule. This means also, as expected, that the associated code distribution method is suited only to short programs. Because transmitting and reassembling long code introduces much overhead than short one.

Note that the purpose of this test is only to compare the processing time at the same node but with different code distribution mechanisms. In these tests, since an EE instance already exists for the first capsule when it passes by the intermediate node for the the second time and for the subsequent capsules, the processing time is much smaller than that for the first capsule traversing a node for the first time. Moreover, processing time at the destination node is much greater than that at the intermediate node. The detailed explanation can be found in section 7.2.1.

6.4   Related Work

In this section, we discuss the related work in the context of AN node architectures and the resource control and accounting technique for the Java programming language.

6.4.1   Active Node Architectures

Since the emergence of the active networking idea, several AN architectures have been implemented, such as ANTS [WGT98], Janos [Janos], CANE [BCZ98], SwitchWare [AAH+98], and Snow [SGW+02] etc., in order to make the network nodes programmable.   As discussed in chapter 2, in general, these projects emphasis on different aspects of AN techniques, such as the level of programmability, security, the speed of processing, the interaction with the legacy systems and the way of service introduction etc.; they use different techniques to make the network active, such as the integrated or discrete approach; and implement different levels of flexibility and security. Compared with these, the most important feature of our AN node architecture is that we have an explicit resource management subsystem in the NodeOS, which provides the resource admission control function, implements resource adaptation, and controls the execution of active applications according to the result of admission control and adaptation, consuming the permitted resources. The main goal of our AN node architecture is to keep the system resource in balance in order to be able to make full use of these resources to serve the active applications.

6.4.2   Resource Accounting and Control in Java

Java has been widely used in the current Internet and in some mobile agent systems for its support of mobility. Correspondingly, resource accounting and control have also been studied for the Java programming language for the sake of security issues.

Jres [CE98] is a work aiming to establish a resource accounting interface for Java. It takes also the CPU, memory and network resource into account, and counts the resources through both the bytecode rewriting and native methods. Basically, we have adopted the same approach to count resources as Jres. The main differences between Jres and our method lie in the following aspects:

First, the goal of Jres is to add a resource accounting interface to Java; whereas our goal is to manage the resources in an AN node. Therefore, Jres counts the resources in a much finer and lower level than we do. Jres counts per thread resource usage; while the resource sharing and accounting unit in our system is an EE instance. For example, for network bandwidth, Jres rewrites the java.net package to locate where a packet is sent and counts it to the corresponding thread, but we can locate and count the packets sent through the NodeOS at the source code level. In addition, we have to consider the synchronization of the counters because of the possibility that multiple threads may access it simultaneously, but this is not a problem for Jres. Moreover, using an EE instance as the accounting unit avoids also the limitation in Jres that the sharing of objects is not handled. Normally threads sharing the same objects belong always to the same EE instance. Generally, the accounting based on an EE instance in our ANwithARM introduces less overhead than that based on each Java thread in Jres.

Second, for the memory resource, Jres counts only the heap used by the Java objects. It does not care about the stack memory. We count also the memory used by the stacks to each EE instance. Similar to Jres, we have not considered the CPU cost for the garbage collection either.

Third, Jres only gives several examples on handling the resource overuse, such as warning and terminating the corresponding threads, lowering the priority of the related threads or adding extra CPU time to these threads; whereas we have a resource scheduling module in the ANwithARM system. The resource overuse processing provides necessary information to the scheduling program.

Another project involving resource accounting in Java is J-SEAL2 [BHV00]. J-SEAL2 is the kernel of a mobile agent system [Bin99], whose design and implementation reconcile strong security with portability and high performance. It is a microkernel design implemented in pure Java and can run on every Java 2 implementation. One of the goals of J-SEAL2 is to support as many different hardware platforms and operating systems as possible, therefore it relies neither on native code nor on modification to the Java Virtual Machine. J-SEAL2 stresses strong portability. It uses only the bytecode rewriting method to count the CPU and memory resource. From this aspect, the resource control in J-SEAL2 is more portable than our system. However, inserting bytecodes for accounting introduces also much overhead. The unit for resource sharing in J-SEAL2 is the protection domain[8], which is the boundary around a component, such as a mobile agent, a service or a set of trust services. To some extend, our EE instance can be seen as one type of such protection domain. However, as a resource accounting unit, our EE instance is relatively fixed, but the protection domain in J-SEAL2 may have various meanings. Furthermore, [Bin01] adopts a benchmark program to count the CPU used for the garbage collection. Namely the J-SEAL2 administrator defines a rough approximation of the number of bytecode instructions required to reclaim an object. By the time that an object is allocated, the

---

[8]J-SEAL2 uses the term of domain as the unit for resource sharing. This should not be confused with the domain in the active node architecture.

bytecode number that will be used for garbage collection is counted. In other words, a protection domain has to pay for the garbage it eventually produces at the time it creates an object. This approach has the advantage that a protection domain is charged for all the garbage it produces, even though the domain is terminated when the related objects are reclaimed. However, the benchmark program includes only the time used to reclaim the objects, the time spent for locating the garbage objects and dealing with the fragmentation is not considered. This method charges in fact only part of the time for garbage collection to the entities.

Profiler is another category of methods regarding the resource usage in Java. Profilers intend to gather information about the resource usage of the Java programs. However, they are designed to help developers optimize the efficiency of their programs, rather than to externally control their resource consumption. The Java Virtual Machine Profiling Interface (JVMPI) [Sun] is an API implemented by Sun, which is in fact a set of hooks to JVM that signals interesting events like thread initiation and object allocations. A profiling tool based on JVMPI can obtain a variety of information such as heavy memory allocation sites, CPU usage hot-spots, unnecessary object retention, and monitor contention, for a comprehensive performance analysis. JVMPI uses indeed the native code support to obtain the resource usage information. Compared with this work, our system uses also the native method to count the resources used by the Java programs. Since the JVMPI is an experimental interface, it is not yet a standard profiling interface, we have used the bytecode rewriting technique to insert our own signals for identifying the thread starting and object allocation.

Chapter 7

System Evaluation

The previous chapters have introduced the adaptive admission control algorithm and the AN node system in which the algorithm is implemented. Furthermore, the major overhead introduced by the key components and algorithms have been evaluated and discussed.

This chapter gives an overall evaluation of the implemented AN node system with adaptive resource management mechanism (ANwithARM). Here we present the results of experiments carried out on the test network consisting of such active nodes. Aspects of the node system are considered in terms of the overhead, the resource utilization and the effects of adaptations. The remainder of this chapter is organized as follows: first the test context is explained briefly, then the functions and implementations of the test applications and a tool that can be used to monitor the resource status in the system are introduced. Following this, the test scenarios and results obtained using these applications and the resource monitoring tool are presented. Finally analyses and discussions of the results are given.

7.1 Evaluation Context

7.1.1 Test Content

We have presented an adaptive resource admission control mechanism in AN nodes, in order to improve the resource utilization of the whole node system, and therefore to increase the probability that an application is accepted and served by a node. This mechanism is implemented in our node system ANwithARM through the integrated resource management subsystem. Generally, the overhead caused by introducing the admission control mechanism in the AN node system involves three aspects. One is the extra processing time. The resource admission control and adaptation needs some extra processing; this may cause some extra delay to the applications of the AN nodes. The second aspect is the adaptive admission control algorithm itself costs also some resources, these are counted to the node system. And finally, some bandwidth overhead may also be caused for the applications because of the transmission of the resource requirement and adaptability information.

In chapter 5 and 6, we have evaluated some components of the ANwithARM node system, such as the optimization algorithm, network bandwidth overhead and the resource accounting. In this chapter, we evaluate the overall node system in terms of the improvement of the resource utilization of the system, the extra delay caused to the active applications that run in the system, and the total resource cost of the system, as well as the effects of the adaptation on a single active application of the system.

7.1.2   Test Network

The test network consists of machines with Intel Pentium(R) III mobile CPU running at 650MHz and 866 MHz and Intel Pentium(R) IV CPU running at 2.0GHz. They are located in the University of Karlsruhe and the Technical University of Braunschweig. All of them are locally connected vice 100Mb/s Ethernet. The network between the universities is provided by the German research network GWIN. These machines run our implementation of the active node system ANwithARM, and construct an overlay active network on the Internet.

7.1.3   Test Means

We adopt two means to evaluate the total ANwithARM system: a resource monitoring tool and some test active applications. Through the resource monitoring tool, the system resource costs and the application resource usage can be observed graphically. And by running the test applications in the AN consisting of the ANwithARM nodes, application related performance data, such as the delay etc. can be obtained. Both means will be described in the next sections.

7.1.4   Resource Monitoring Tool

As mentioned in section 6.1.3, we have implemented a resource monitoring tool in the active application (AA) part of the node system, to monitor the various resource usage in the node system.

The resource monitoring tool is implemented as a pure active application, called ResMonitor. That means, when it is started, an EE instance is created for it and certain resources are allocated to this EE instance, just like the normal active applications from end-users. It can also invoke all the services provided by the underlying EE and NodeOS. The principle of this tool is that ResMonitor polls the various resource usage information from the node system periodically and then displays the obtained data graphically. Therefore, functionally the application consists of two parts: data query and display.

The data query part acquires the needed resource usage information through invoking functions provided by the NodeOS, because only the NodeOS holds the diverse resource status information in the system. However, in order to reduce the time and resource cost incurred by invoking the functions between the AA and EE as well as between EE and NodeOS, the resource monitoring application, as an embedded system application, is given the privilege to be able to invoke the resource query functions directly instead of being required to go via EE like the normal applications from end-users must do.

In our current implementation and configuration, the NodeOS functions osResUsed(), anNodeResUsed and appsResUsed() are invoked every one second. By this the total resources consumed in the underlying Windows OS, in the ANwithARM system, as well as the resources consumed by each active application running in the ANwithARM node system, are collected and displayed dynamically. However, the polling time interval can be changed according to the need of the monitoring. Note that for the purpose of display, the CPU usage is provided in percentage of the total CPU capacity by the NodeOS functions.

Displaying the data of the resource status graphically is implemented under the Windows XP OS, using Microsoft platform SDK (Microsoft Platform Software Development Kit) [Micro1]. Figure 7.1 illustrates an example of the display window. The instantaneous information of the resource usage of the Windows OS, the whole ANwithARM system and the applications running in the node system is illustrated in the upper part of the window. Users can select the resource type to be observed, e.g., the CPU, memory or network bandwidth, using the index card. So far the instantaneous information about maximal 10 active applications in the system (the top 10 applications running in the node system) can be displayed due to the space reason[1]. In this example, there are three applications running in the ANwithARM: MonitorRemote, VideoServer and ImageServer. In the lower part, the history of the specified type of resource in displayed. And on the bottom of the window, some statistic information is listed.

Moreover, when the index card Applications is selected, information about the applications currently running in the node system is listed, including the name of the applications, the class name of the executed Java bytecode, the execution status, as well as the CPU and memory usage. Similar to the Applications card, when the index card Threads is selected, information about the threads having the type of ThreadWrapper which wraps the main threads created by the application programs running in the node system is listed.

The logical relationship among the different components for implementing the above display window is illustrated in Figure 7.2. Here the index card of CPU, memory and network bandwidth is instanced using JCPUPanel, JMEMPanel and JBWPanel respectively. All these three classes are derived from the parent ChartPanel and have an instance of HistorySurface, which can draw the process of resource usage. The index card Applications and Threads are initiated by the JAPPPanel and the JThreadPanel, which have the common parent JTablePanel. Therefore they both have the commonality like having a

---

[1]Information about all the applications running in the system can be inquired using the index card Applications.

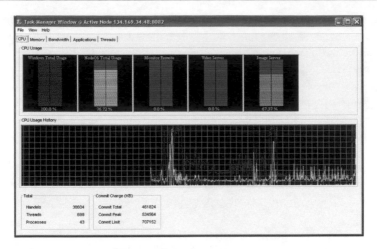

Figure 7.1: Resource monitoring tool

dynamic table, a ButtonPanel etc. All these components are controlled by the TaskMan-
agerWindow, which is started according to the need of users during the initiation of the
ANwithARM node system.

In addition, the colors of each display panel can also be selected from the menu File in
the main display window, and the dynamically displayed resource usage curve can also
be stored for the purpose of later analyses or statistics. Currently, the displayed curve
can be stored in three formats. One is graph, namely the resource usage history can be
saved directly as GIF or JPEG file. Another is XML text, i.e., the resource data can also
be saved in the form of XML text. For this purpose, a DTD (Document Type Definition)
file is defined to regulate the structure and the resource usage data. The third format
is Java Object. Namely the displayed curve can be saved as Java Objects through the
serialization procedure, so that the data can be used directly by some data analysis tool,
such as JFreeChart.

We have also extended the function of this tool, so that the resource usage status
at a remote active node can also be observed. The principle is to collect the needed
resource status data at the specified remote AN node, transfer them back using the active
application technique, and then display the obtained resource data locally. I.e., users
program their monitor demand and send packets carrying the program to the remote
AN node which should be monitored. There the program is loaded into the EE and
executed. During the execution, the resource query functions provided by the NodeOS,
as introduced in section 6.1.1, are invoked and the desired resource information are
acquired. The obtained data is then packed into packets and sent back to the monitoring
node.

Figure 7.3 illustrates this principle and the types of packets used. The MonitorStart

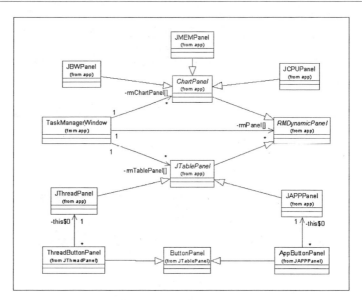

Figure 7.2: UML-Class diagram for the main display window

packet carries the program. It tells the remote node what kind of resource status information is needed and how to acquire it. The ResourceData packets carry the collected resource data back to the user continuously. The other packets are control packets. The MonitorSuspend and the MonitorResume packet can pause the remote monitoring procedure for some time. The FrequencyChange packet can change the frequency with which the ResourceData packets are sent back to the monitoring node. And the MonitorStop packet can stop the remote monitoring procedure by ceasing the execution of the program at the remote node.

As the resource status data at the remote node are transferred to the local monitoring node through the ResourceData packets, they are displayed using the same method as introduced above. In conclusion, the only difference between the remote and the local resource monitoring is that the source of the data displayed is different: the former acquires the data through unpacking the ResourceData packets, and the later acquires the data by polling the local node system directly. More details about the remote monitoring method can be found in [LWX03].

7.1.5   Test Applications

Besides the resource monitoring tool, we also developed the Traceroute and the Image application on top of the ANwithARM node system. The functionality provided by these two applications is simple. Moreover, they are not intended to be of significant practical

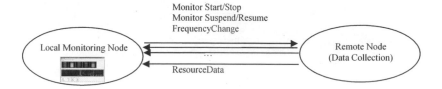

Figure 7.3: Principle of remote resource monitoring

use in real ANs. However, their main purpose is to be suitable for testing the AN node system with the adaptive admission control mechanism and get the related performance data. Both applications are adaptable and may, hence, consume different amount of resources. The packets of the two applications have the general format as depicted in Figure 6.5, and they carry the desired resource requirement and the ARVS information in the ANEP header.

7.1.5.1 Traceroute Application

The Traceroute application is implemented mainly to observe the processing delay and the resource costs caused by the ANwithARM node system. Since the application itself does not require any complex functional processing in the active nodes, except for recording some specified information in the specified active nodes, we can use it to observe the resource usage of ANwithARM, and use the delay that the Traceroute application experiences at an AN node to approximate the extra delay exerted on a normal active application by the ANwithARM node system.

Similar to the Traceroute program in the traditional IP network, our Traceroute application can also be used to obtain the route and transmission delay information between a source and a destination by sending packets to the destination node. However, unlike the IP Traceroute program, which uses the ICMP message to get the route and delay information, our Traceroute application obtains the information actively through the TracerouteCapsules. It can specify where and what kind of information should be reported in the TracerouteCapsules. The most important difference between our Traceroute application and the Traceroute program in the traditional IP network lies in two aspects. One is that more information about the traversing network nodes, e.g., the resource usage information, can be obtained by our Traceroute application with the help of the underlying AN. The other is that users can specify the node, i.e., the concerned node, at which the information should be collected; whereas information at others nodes on the way from the source to the destination is neglected. In our current implementation, the parameters that can be specified by the users include the concerned node addresses, the number of the TracerouteCapsules to be sent, and the interval for sending the TracerouteCapsules. If no concerned node address is specified, the information about all the active nodes that the TracerouteCapsule passes through is collected.

Figure 7.4 illustrates the format of the TracerouteCapsule passing through the AN-withARM nodes. The desired resource requirement r_d and the ARVS information are packed in the ANEP header of the first TracerouteCapsule sent to the network, as mentioned in the last chapters. Moreover, a ConcernedNode option field is also put in the ANEP header of the capsules, so that the node system can check quickly as a Traceroute-Capsule arrives at the node, if this node is the concerned node. If not, the node forwards the capsule immediately, and if yes, the specified information is recorded in the capsule and the capsule is then forwarded continuously. Note that the ConcernedNode option field has also the structure of TLV (Type/Length/Value), just like other options introduced in section 6.1.2. The value field consists of one or several node addresses, at which the specified information should be collected. A detail description about the format of this option can be found in appendix A.

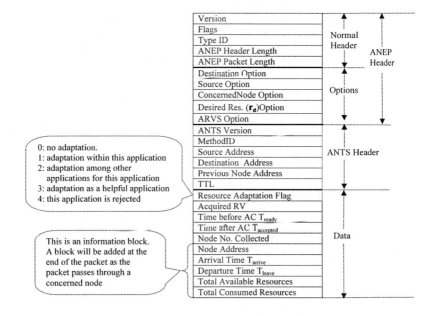

Figure 7.4: The content of the TracerouteCapsule

The Information that the Traceroute application can collect at each concerned node are:

- The time when the capsule arrives at the node.

- The time just before the packet leaves the node.

- The total available resources at the node.

- The total resources consumed by the ANwithARM, including resources used by the node system and by the active applications running in the node system.

Among them, the resource information is obtained through invoking the resource query functions provided by the NodeOS through the EE. All the information about an AN node is organized in an information block. As a TracerouteCapsule passes through an AN node, if the address of this node is specified in the ConcernedNode option field, or there is no ConcernedNode option field in the packet header, an information block is added into the TracerouteCapsule and the capsule is forwarded continuously. Note that on the way from a source to a destination and from the destination back to the source, a TracerouteCapsule may pass through some intermediate nodes twice. In this case, information about these nodes is collected twice. In addition, in our implementation, the destination node is always the concerned node.

Moreover, information about the resource admission control at the destination node is also recorded in the TracerouteCapsule. We have defined five identifier members to indicate whether and what kind of adaptations occurred during the admission control related to the application at the destination node. Namely

0: No adaptation occurred for this application during the admission control;

1: Adaptation within this application has happened before the application is accepted;

2: This application is accepted only after adaptations among other applications in the node;

3: During the running of the application, adaptation within this application has occurred for the sake of helping others;

4: There are not enough resources at the destination node for this application; thus, the following TracerouteCapsules belonging to this application will not be processed by the destination node. Logically, when a TracerouteCapsule with such an adaptation flag returns to the source, the application will stop sending the TracerouteCapsules immediately if there are still TracerouteCapsules not yet sent.

Moreover, the time point right before and after the admission control, as well as the result of the admission control, namely the resources acquired by this application, are also recorded and carried back to the source node by the TracerouteCapsules.

For the tests in this chapter, we have adopted the on-demand code loading mechanism to distribute the code for processing the TracerouteCapsules, since this mechanism does not require the first TracerouteCapsule to carry the code for processing itself. Moreover, in case the needed code is already cached in a network node, this mechanism can also eliminate the processing delay caused by the code distribution. This is helpful for us to

```
//the code belongs to the Class TracerouteCapsule, which is distributed to an AN node on-demand
public boolean evaluate(Node n) { //n is the node where the packet arrives
   if (n.address()!=getSrc() ){
      nodePassed ++;

      if (n.address() == getDst()) {
         arrived=true;
         int acFlag=n.deliverToApp(this);
         if (acFlag==4) // the creation of app is rejected, notify the source
            n.forward(getSrc(),4);
      }
         //decide if info at the current node should be collected
         //and then forward itself continuously.
      else {
         it ( n.address()== concernedNodeAddr() )
            collectNodeInfo(n);
         if (arrived!=true) return n.forward(getDst());
         else return n.forward(getSrc());
      }
   }
   else //the capsule returns to the source node, deliver to the TracerouteApp to
        //extract the carried info
      return n.deliverToApp();
}


//the following code belongs to the class Node, which belongs to the AN node system
public int deliverToApp(Capsule cap) {
   boolean newApp=decideNewApp(cap); //decide if it is a new app
   if (newApp==true){
      int returnValue=Sys.admissionControl();
      tAccept=getCurrentTime(); //record time right after admission control
      if (returnValue!=4) { //the new app can be accepted.
         createApp (cap, res, tAccept,returnValue); //res is from the admission control
            // create a new app in the node system with the allocated resource
      }
      else //not enough res for this app, i.e., cap.acquiredRV=0;
   }
   else {
      int returnValue=Sys.ACMark(); //get the adaptation mark related to this app.
      deliverApp(cap); //deliver the app to the already exist app
   }
   return retunValue;
}
```

Figure 7.5: Pseudo program code for processing the TracerouteCapsule

measure the processing overhead of ANwithARM from the point of view of an active application. Figure 7.5 illustrates the pseudo code for processing the TracerouteCapsules at the active nodes. To keep the generality, an application level EE instance is created at the destination node. The tasks of the created application at the destination node includes allocating some objects and adding some pads to consume the allocated memory, bandwidth resource and part of the CPU resource, recording the admission control result information and the time as well as some resource usage information. Note that due to the creation of the new EE instance, a new TracerouteCapsule is spawned at the destination node and sent back to the source instead of the original one. In addition, the code for obtaining the arrival and departure time are not illustrated. In order to achieve more accuracy, we record the arrival time right after a capsule is received and begin to be unpacked[2]. The departure time is recorded just before it is packed into the capsule and then the capsule is sent out. In addition, the admission control is configured to be performed at the "application-level". Namely at the intermediate nodes, only the packet forwarding is executed, no admission control procedure is performed. Note that we have adopted the general methods for developing active applications using ANTS EE [WGT98] which is the basis of our EE.

7.1.5.2 Image Application

The Image application allows users to fetch a specific image from a remote image server. The obtained image may have different qualities depending on the resource status at the remote server when the image is requested, and on the ARVS information of the user applications, which represents the tolerance of users to the image qualities. It can be used to determine the amount of the resources acquired by an active application after the admission control and to demonstrate the amount of resources that the application actually consumed under the scheduling of the resource management subsystem in the node system. Another goal for implementing this application is to give an example of the adaptive applications in our active node system, including why and how to utilize the application adaptability.

The general assumption of this application is that the images are stored in one format only, i.e., as PNG files, in the image servers. Users want to get the images in the original PNG format in the ideal case. Nevertheless, they can also accept images in another format or compressed to some specified degree. Due to the resource limitations in the server node system, the servers cannot always send the image with the ideal quality to the users. Therefore some processing on the image data is performed in the server according to the amount of resources allocated to the users after the admission control. Then the images with the corresponding qualities are sent to the users. In other words, the users can get the image with the "best" quality through making full use of the current amount of resources that they can acquire at the server.

The Image application implements the ImageProtocol, which can process two types of capsules: the ImageRequestCapsule and the ImageResponseCapsule. The ImageRequest-Capsule is initiated by the user, indicating the name of the requested image, together with the desired resource requirement r_d, the ARVS information and the specified server node. Here r_d and the ARVS reflect implicitly the desired and the acceptable quality of the image. The ImageResponseCapsule is used to carry the data of the obtained image from the specified server node back to the user requesting the image. Considering the size of the image file and the limit of the UDP packet length[3], the image data is sent back to the client in several ImageResponseCapsules. Besides the image data, the name, format

---

[2]At this moment, it is still not known if the received capsule is a TracerouteCapsule.

[3]For our ANwithARM, the ANEP packet is finally packed into a UDP datagram, which is then attached with IP headers and transferred through the physical network. Therefore, the maximum permitted capsule length is in fact limited by the length of the UDP packet payload.

and compression level (if the image has been compressed) of the image, as well as the segment number of the image data are also carried in each ImageResponseCapsule. After all the image data segments are received at the client, the image file is reassembled, decompressed if necessary, and displayed. Additionally, the IP address of the server, the image format and the compression level, if there is any compression, are also presented.

When an ImageRequestCapsule arrives at the specified server, the server node system first performs the adaptive admission control according to the r_d and ARVS carried in the capsule and the available resources in the node system, to decide how many resources can be allocated to this image request application. Then the data of the specified image is read from the file system and processed according to the amount of resources allocated to this application. The processing on the source image file may include:

- Select a suitable image format.

- Scale the image to a suitable size.

- Compress the image to a suitable level.

To establish the relationship between the processing and the resources needed, we used a mapping table for each image in the server nodes, indicating how many resources are needed for each type of image format. E.g., to convert the original PNG file to JPEG with the compression level of 90, this mapping table states how many bytecodes need to be executed, how much memory is needed and how long the resulting file is (which affects the network bandwidth). This information has been achieved through pre-processing each image in advance. The table contains information for converting the PNG file with original size, half and quarter size (both image width and height) to JPEG with compression quality from 10 to 100 at an interval of 10, as well as the information about the original PNG file.

Figure 7.6 shows the pseudo code for processing the ImageRequestCapsule in the server system after the admission control. quality=−1 means the original PNG file can be transmitted, otherwise the image has to be converted to JPEG format. quality=10~100 means the size of the image does not need to be changed, but the image should be compressed at the corresponding level. quality=210~300 means first the image has to be reduced to the half size, and then compressed at the corresponding level. Similarly, for quality=410~500, the image size has to be reduced to the quarter and then compressed at the specified level. After the image has been reduced and compressed, the image data is sent back to the user in several ImageResponseCapsules, depending on the final length of the image having been adjusted and the maximum length of each capsule permitted by the underlying active network. Note that all the resources consumed in the processing, from determining the image quality to sending data to the users, are charged to the corresponding user application.

Through this method, users may get images with different qualities depending on the resource status in the server system and the amount of resources allocated to it. In other words, user applications can adapt themselves to the resource conditions in the server node system.

```
public void receive(ImageRequestCapsule imCap, RV acqRes) {
    //acqRes gives the acquired resource passed by the RM module after the admission control

    int quality=determineImageQuality (acqRes, imCap.name);

    if ( quality==-1 ) { //the resource is sufficient for the original image
        imName=this.getNodeAddress()+ ":" +imCap.name();
        //add the server's IP address before the request image
        fis =new FileInputStream(IMAGELOCATION+"/"+imCap.name);
        imLen=fis.read(imData);
        //the image data is read to imData, and the length is stored in imLen;
    }
    else { //the image has to be compressed or reduced in terms of the specified quality
        imName=this.getNodeAddress()+ ":" +imCap.nameWithoutExt()+".jpg" + "-quality "+quality;
        //add also the compression level
        int rr=quality/100;  //reduce rate of the image size
        comLevel=quality%100; //compression level of the JPEG
        Image imageOrg = toolkit.getImage(IMAGELOCATION+"/"+imCap.name);
        Image image=imageOrg.getScaledInstance(imageOrg.width/rr,imageOrg.height/rr);
        //first reduce the size of the image, then compress
        JpegEncoder jpeg= new JpegEncoder(image,comLevel,(OutputStream)out);
        jpeg.Compress();
        imData =out.toByteArray(); //the compressed jpeg image data is now stored in ImData
        imLen=out.size();
    }
    sendImageResponseCapsule(destination,imageData,imLen);
        // destination is equal to the source address of the ImageRequestcapsule
}

public void sendImageResponseCapsule(Int dest, byte[] imData, int imLen) {
    int dataChunks = (imLen +CHUNK_SIZE -1)/CHUNK_SIZE;
        //image is sent in several capsules
    index=0;  //index of the chunk
    for (int p=0; p<imLen; p+=CHUNK_SIZE) {
    ImageResponseCapsule respCap= new ImageResponseCapsule(imName, imLen, imData,
                                    dataChunks,idx,p,l);
                            //l=imLen-p, considering the last segment
    boolean sent =thisNode().routeForNode(respCap, dest);
}
```

Figure 7.6: Pseudo code for processing the ImageRequestCapsule at image server

Here we have only implemented the Image application following the simple client/ server pattern, i.e., the acquired image is sent back from the server to the user who has requested it, and no further processing within the networks is done. In practice, this pattern can be extended. E.g., the acquired image can be sent to another user. In this case, through the adaptive admission control, the server node acts as a transcoding proxy: the receiver gets the image according to the resource conditions in the network node. If the admission control mechanism can also be extended, and the server nodes know some information about the receivers, e.g., the device type or the resource capability such as the processing power and network capability etc., the server nodes can also act as the filter and realize different scaling functions according to the features of the applications and the capability of the receivers.

7.2   Test Scenarios and Results

This section presents some results of the experiments carried out on the test network consisting of our ANwithARM active node systems. The experiments are performed in terms of measuring the overhead introduced by the admission control, including the resource costs and the delay caused by the adaptive admission control procedure, and the improvement of the resource utilization of the total node system through the adaptations, as well as the possible effects of the adaptations on the active applications.

7.2.1   Processing Time and Resource Costs

As mentioned above, we can use the Traceroute application to observe the delay and resource costs caused by ANwithARM. On the one hand, the Traceroute application itself does not perform any complex actions in the destination node system, except for recording some information about time, resource usage and resource adaptations. Hence, the time that the TracerouteCapsules delay at the destination node can be approximately regarded as the processing delay caused by the whole ANwithARM node system from the point of view of active applications. By observing and comparing the resource consumption of ANwithARM under different situations, we can also achieve a general perception about the resource costs of ANwithARM and the adaptive admission control procedure. On the other hand, since the Traceroute application itself does not need extra resources for performing complicated tasks, it is relatively easy to control the desired resource requirement and the ARVS information of this application, and to simulate the admission control in different cases, such as adaptation within an application or among different applications.

Since each TracerouteCapsule records the time when it arrives at the destination node t_{arrive}, when the admission control begins for it t_{ready}, when it is accepted by the node t_{accept}, and when it leaves the destination node t_{leave}, we use $t_{accept} - t_{ready}$ to denote the time used for admission control, and $t_{leave} - t_{arrive}$ to denote the total processing overhead introduced by ANwithARM in the following tests. In addition, the resource monitoring tool at the destination node is initiated to observe the resources usage there.

7.2.1.1   Test Network Structure

The test network is shown in Figure 7.7. It consists of three active nodes, each runs our implementation of the node system ANwithARM, acting as the source node, intermediate node and destination node respectively. These nodes are configured in such a way that all capsules sent from the source to the destination have to pass through the intermediate node. And these capsules come back from the destination to the source node along the same way as they arrive at the destination. The source and the destination node are located in Karlsruhe, and the intermediate node is in Braunschweig. At the destination node, ANwithARM runs on the machine with the Intel Pentium(R) III mobile CPU running at 866 MHz, and at the source and the intermediate node, ANwithARM runs on the machine with Intel Pentium(R) III mobile CPU running at 650 MHz and Pentium(R) IV CPU running at 2.0 GHZ respectively.

Figure 7.7: Path for the Traceroute application

7.2.1.2 Tests and Results

In the following tests, we control the value of r_d carried in the first TracerouteCapsule and the resource status in the destination node system to control the occurrence of the resource adaptations in the admission control phase and record the processing time under different types of adaptations. Since we concentrated on the processing overhead caused by the adaptive admission control at the destination node, we adopt the on-demand code distribution method and let the code for processing the TracerouteCapsules be cached in advance in the intermediate node, so that the overhead caused by the code distribution can be neglected.

Without adaptation

In the first test, the Traceroute application has the following resource requirement: bandwidth 20000 bytes/s, CPU 300000 bytecodes/s, and memory 40000bytes, denoted as (20000, 300000, 40000). The corresponding lower limits are (12000, 130000, 30000), and the upper limits are (40000, 500000, 6000000). The efficacy function is assumed to be linear, namely $Ef=0.58b + 0.18c + 0.24m$, when (b, c, m) is in the range of the limits explained above. No values for the worst efficacy (WE) and the highest cost (HC) are specified. In this test, 90 TracerouteCapsules are sent to the destination at an interval of 2000ms[4]. Before the first capsule reaches the destination node, the available resources in the system are limited to (28000, 30000000, 69259260), i.e., there are enough resources in the node system for this Traceroute application.

Figure 7.8 illustrates the amount of resources allocated to the Traceroute application at the time when each capsule arrives at the destination node as well as the processing and admission control (AC) time experienced by each capsule at the intermediate and/or the destination node. These values are an average of 10 tests[5]. The value for the capsule number from 3 to 90, is an average for these capsules of the 10 tests. Note that under

---

[4]To keep the generality of the test results and to be able to control the duration for observations, we sent 90 capsules at an interval of 2000ms. This is also applicable for the following tests.

[5]Note that for each test, the available CPU and memory resource in the system have a little difference, but they are both large enough to guarantee that no adaptations will occur.

our network configuration, the capsules pass through the intermediate node twice, i.e., on the way from source to destination and from the destination back to the source. What we have shown in Figure 7.8 is when the capsules pass through the intermediate node for the first time.

Required resource	Capsule No.	Acquired resource	Processing time at intermediate (ms)	Processing time at destination (ms)	Time for AC at destination (ms)	Total transmission delay (ms)
(20000, 300000, 40000)	1	(20000, 300000,40000)	4.667	52.303	11.124	118.367
	2	(20000, 300000,40000)	0.876	3.032	0	60.012
	3-90	(20000, 300000,40000)	0.974	3.352	0	61.235

Figure 7.8: Processing time and total transmission delay: no adaptation occurs

In this test, the Traceroute application got the required resources from the beginning till the end. No adaptation occurs during the admission control and the execution of the application. Here the processing time is calculated according to the arrival and the leaving time recorded in each capsule when it passes the active nodes, namely $t_{leave} - t_{arrive}$. For the first packet, at the intermediate node, the processing time consists of the time used for unpacking the packet, deciding whether a new EE needs to be created, creating a new EE, searching the method for processing the packet in the local cache according to the method identifier (ID) carried in the packet, and executing it in the EE to process the packet. Note that at the intermediate node, only a new protocol level EE instance may be created, as no application-level domain is initiated. Since we have configured the node system to perform admission control at the application level EE instance, no admission control is performed at the intermediate node.

At the destination node, the processing time comprises mainly three parts. One is the same as that in the intermediate node, namely unpacking the packet to find the needed information, deciding whether a new EE instance should be created, and searching the specified method. The second part is that used for admission control. It is calculated through $t_{accept} - t_{ready}$, i.e., the interval between the time when it is determined that a new EE should be created and when it is confirmed that the EE instance with the specified resource requirement can or cannot be created in the system. And the last part is for the creation of an application-level EE instance. If the admission control is passed, an application-level EE instance is created with the allocated resources, and the assigned tasks are executed in it. Since an application-level EE instance needs to be created at the destination node, a new TracerouteCapsule is spawned and sent, and the system must allocate a resource controller (GRC) to control the execution of the application, the processing time for the first packet in the destination node is much larger than that in the intermediate node.

For the subsequent TracerouteCapsules, the nodes only have to pass them to the corresponding EE instance which already exists. At the destination node, similar to the first capsule, the capsules are passed to the application level EE instance, and as a result of the execution of the application, the capsules sent back to the source must be generated, the processing time is longer than that at the intermediate nodes where only some information about the node are inserted and the capsule is forwarded continuously.

The total transmission delay is the interval between the time when a TracerouteCapsule is sent out at the source and when it comes back to the source. It consists of the transmission time from source to intermediate node, and then to the destination and from the destination back to the source passing through the intermediate node again, as well as the processing time at the intermediate nodes and the destination node.

The resource usage of ANwithARM and the Traceroute application observed by the resource monitoring tool at the destination node is illustrated in Figure 7.9. Here the resource usage of ANwithARM includes three parts. One is that used by the system threads for keeping the operation of the node system, such as threads monitoring the arrival of capsules from the network, the threads managing the system resources etc. The second part is that cost by the resource monitoring tool itself. And the last part is the resources cost by the Traceroute application. From this figure we can see that the arrival of an application cause a peak in the system CPU usage (Figure 7.9(b)). And when there is an application running in ANwithARM, its CPU cost is about 7.5%, however, in the idle case, i.e., no active applications from end-users running in the node system, its CPU cost is about 6%.

The Traceroute application costs CPU resource mainly when a TracerouteCapsule comes, and it is in fact used for collecting time and resource information, as well as for the sending and receiving of TracerouteCapsules. Since the ANwithARM node system does not have any bandwidth costs by itself, the bandwidth usage of the node system is the same as the bandwidth usage of the Traceroute application (Figure 7.9(a)), and therefore is not depicted. In Figure 7.9(a), the ANwithARM Available means the available, i.e., the limited bandwidth in the system before the first TracerouteCapsule came. For the memory resource, the total system memory status varied more strongly when the Traceroute application ran in the node system than in the idle case, i.e., no application running in the system (Figure 7.9(c)). This is because the garbage collection is performed more often during the existing of the GRC for the Traceroute application than in the idle case.

Adaptation within one application

In the second test, the Traceroute application has the same resource request and ARVS as in the first one. But the available resources at the destination node are limited, i.e., we limited the network bandwidth to 35000 bytes/s. Similar to the first test, we repeat the test for several times to get an average value for the processing time and the total transmission delay; and each time, the available CPU and memory resource may differ slightly, but both are large enough for the Traceroute application. For the first time of the test, the available resource in the node system is (35000, 30000000,69922432), i.e., as the first TracerouteCapsule arrives at the node. Figure 7.10 shows the result of this test.

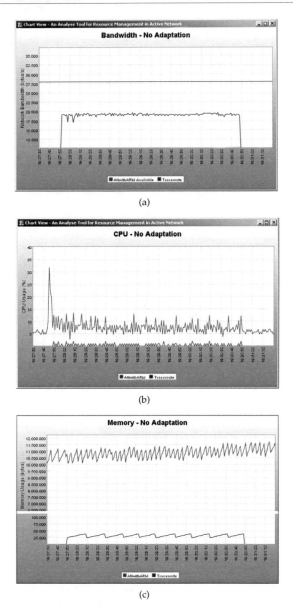

Figure 7.9: Resource usage at the destination node: no adaptation occurs

From the result we can see that adaptation occurs at the destination node: the application got less bandwidth, but more memory and CPU resource than required. The

Required Res.	Capsule No.	Acquired Res.	Proc. time (ms) at inter.	Proc. time (ms) at dest.	Time for AC at dest.	Total trans. delay (ms)
(40000, 300000, 40000)	1	(25030,264636,84648)	4.862	58.012	15.355	125.122
	2	(25030,264636,84648)	0.897	3.253	0	60.562
	3-90	(25030,264636,84648)	1.132	3.533	0	60.972

Figure 7.10: Processing time and total transmission delay: adaptation within an application

processing time at the intermediate nodes has almost not change; however, the process-ing time at the destination node has increased due to the increase of the time for the admission control. Note that in Figure 7.10, the acquired resources are randomly selected from one test, since for each test, the results may differ slightly due to the difference of the available memory and CPU resources in the system and the selection of the initial values of the Simplex optimization method. Other values are average values from 10 tests, just like in the first test.

Figure 7.11 illustrates the resource usage in this case. This time the peak in the CPU usage (Figure 7.11(b)) caused by the arrival of the application is greater than that shown in Figure 7.9(b). This is because the system has also to spent some extra resource on the admission control due to the occurrence of the resource adaptation besides preparing an environment for the application execution. In addition, during the execution of the application in the system, there is no obvious difference for the CPU cost of the ANwithARM system compared with that in the first test. Nevertheless, the CPU cost of the Traceroute application becomes a little smaller due to the adaptation, but its memory and bandwidth consumption (Figure 7.11(c) and Figure 7.11(a)) is obviously larger than that in the first test.

Adaptation among applications

In the third test, we started two Traceroute applications with an duration of 1 minute and 30s between them. Both applications had the same resource requirements and ARVS parameters, except for RV_L, as summarized in Figure 7.12. The system available resources before the first TracerouteCapsule of the first application arrived were (35000, 30000000, 196134252), i.e., the available bandwidth was not sufficient when the two applications run simultaneously in the node system. We study the resource adaptation at the destination node through the amount of resources each application acquires. Figure 7.12 illustrates the results of this test. Since there is not any variation for the processing time at the intermediate node form the perspective of both theory and practice, we have discarded this item in Figure 7.12.

From this figure we can see that for the first Traceroute application, from the 46th capsule, i.e., after the second application arrived, the resources it acquired changed, whereas the second application obtained the required resources and this amount did not

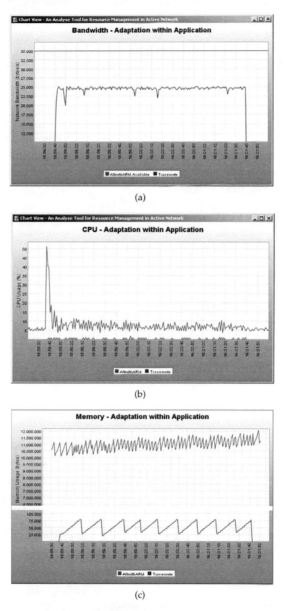

Figure 7.11: Resource usage at destination node: adaptation within an application

change after the first application stops. This is because there was not enough available bandwidth in the node system when the second application arrived. Due to the relative

	Cap. No.	Acquired RV	Proc. time (ms) at dest.	Time (ms) for AC	Total tran. delay (ms)
Application 1	1	(20000,300000,40000)	52.497	11.433	118.072
r=(20000,300000,40000)	2	(20000,300000,40000)	3.113	0	61.274
RV$_L$=(12000,130000,30000)	3~45	(20000,300000,40000)	3.251	0	61.274
RV$_H$=(40000,500000,6000000)	46	(13382,130001,175496)	3.379	0	61.732
	47~90	(13382,130001,175496)	3.301	0	61.612
Application 2	1	(20000,300000,40000)	60.912	18.591	126.963
r=(20000,300000,40000)	2	(20000,300000,40000)	3.345	0	61.727
RV$_L$=(16000,130000,30000)	3~45	(20000,300000,40000)	3.313	0	61.699
RV$_H$=(40000,500000,6000000)	46~90	(20000,300000,40000)	3.310	0	61.402

Figure 7.12: Processing time and total transmission delay: adaptation among applications

higher bandwidth limitation of the **RV$_L$** of the second application, the adaptation inside it could not succeed. Therefore, the first application became the helpful application and the adaptation within it was successful. As a result, both the CPU and network bandwidth resource allocated to the first application were lowered, and the memory allocated to it was increased; whereas the second application obtained the required resources from the beginning to the end. Furthermore, compared with the second test, the processing time at the node and the total transmission delay are slightly increased. This is because two optimization procedures have occurred in the third test.

Figure 7.13 illustrates the resource usage of the two Traceroute applications and the ANwithARM system observed by the resource monitoring tool. Here we can see that the arrival of the second application caused a peak in the CPU usage of the whole node system due to the admission control and the initiation of a new application in the node system, and the first application consumed a little fewer CPU and bandwidth resource when the second application began to run. However, the memory usage of the first application increased obviously, whereas the bandwidth usage decreased, until it ended. The resource usage of the second application had no change. From the beginning to the end, it acquired the required resource. Similar to the second test, the system memory usage varies more strongly when two applications ran concurrently compared with that when only one application was running in the system.

Without admission control

To examine the performance of the ANwithARM system, we have also implemented a node system using the ANTS2.0 and Jnodeos but without the resource management subsystem. We run the same Traceroute application again in this node system. At this time all the nodes have enough resources to let the packets pass through. Figure 7.14 illustrates the processing time at the node system and the total transmission delay of the capsules. Similarly to the first test, the values are averages of 10 times, and for capsule

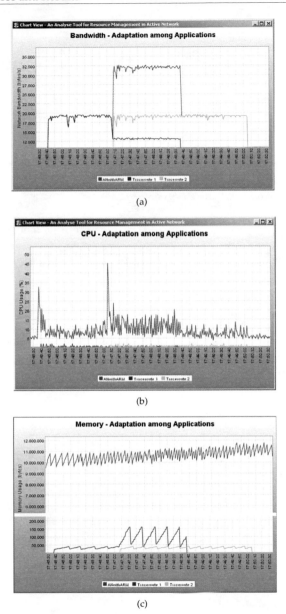

Figure 7.13: Resource usage at destination node: adaptation among applications

number 3-90, the values are an average of the values of all these capsules.

Capsule No.	Proc. time at inter. (ms)	Proc. time at dest. (ms)	Total trans. delay (ms)
1	3.052	30.092	92.124
2	0.633	2.977	60.533
3-90	0.667	2.957	60.921

Figure 7.14: Results without admission control

Together with Figure 7.8, we can see that when the resource management subsystem is integrated, about 20ms processing overhead is introduced for the first capsule and nearly none for the subsequent capsules. Among them, about 11ms is spent on admission control.

In order to have a general view about the basic transmission time between the node systems located in Karlsruhe and Braunschweig, we list the route and the transmission time of the underlying IP network between Karlsruhe and Braunschweig acquired using the traditional Traceroute program in Figure 7.15.

In conclusion, the ANwithARM node system can adjust the resources allocated to the applications according to the resource status in the system, and the applications consume the acquired resource under the control of the node system. The admission control algorithm itself costs some system resources and may also cause some processing delay; and for different types of adaptation, the amount of the resource costs and the processing time have also some difference. However, they are both acceptable.

7.2.2   Resource Utilization

The above section has evaluated the processing overhead of the ANwithARM AN node system. In this section, we observe the changes of the system resource utilization caused by the resource admission control mechanism in ANwithARM by redistributing resources among different applications.

For this purpose, we execute applications with different resource requirements and ARVS and observe if they can be accepted by AN nodes and how many resources they can acquire under different system resource conditions. At the same time, we check if any and what kinds of adaptations happen and calculate the corresponding utilization of the different resource types.

7.2.2.1   Network Configuration

The network configuration and the applications running in this test are depicted in Figure 7.16. Here four AN nodes are involved. Furthermore, we have not used any intermediate AN nodes[6], in order to be able to control the occurrence of the resource adaptations in the

---

[6]No intermediate active node means that the active packets (i.e., capsules) are not passed and forwarded by another active node.

```
traceroute to 134.169.34.130 (134.169.34.130), 64 hops max, 40 byte packets
 1  i70gate.tm.uni-karlsruhe.de (141.3.67.254)  0.251 ms  0.125 ms  0.117 ms
 2  172.16.4.1 (172.16.4.1)  0.884 ms  0.723 ms  0.348 ms
 3  192.168.1.190 (192.168.1.190)  2.973 ms  0.946 ms  0.953
 4  172.21.3.9 (172.21.3.9)  18.673 ms  0.885 ms  0.613 ms
 5  Karlsruhe1.BelWue.de (129.143.166.129)  0.792 ms  0.917 ms  0.655 ms
 6  Stuttgart1.belwue.de (129.143.1.7)  2.932 ms  1.875 ms  1.889 ms
 7  ar-stuttgart2-ge6-0-0.g-win.dfn.de (188.1.38.53)  2.932 ms  2.025 ms  3.214 ms
 8  cr-stuttgart1-ge5-0.g-win.dfn.de (188.1.76.1)  3.342 ms  2.978 ms  4.194 ms
 9  cr-frankfurt1-po8-0.g-win.dfn.de (188.1.18.77) [MPLS: Label 12508 Exp 0]  11.194 ms
       1.910 ms  11.872 ms
10  cr-hannover1-po3-1.g-win.dfn.de (188.1.18.182)  11.272 ms  11.301 ms  10.247 ms
11  ar-braunschweig3-po5-0.g-win.dfn.de (188.1.88.70)  15.677 ms  13.233 ms  14.499 ms
12  ciscobsw.rz.tu-bs.de (134.169.3.222)  20.637 ms  13.784 ms  14.992 ms
13  ibrgate.rz.tu-bs.de (134.169.246.34)  13.618 ms  14.233 ms  13.784 ms
14  ibrgate.rz.tu-bs.de (134.169.246.34)  14.667 ms  14.062 ms  15.001 ms
15  ibrgate.rz.tu-bs.de (134.169.246.34)  14.805 ms  14.213 ms  14.791 ms
```

(a) From Karlsruhe to Braunschweig

```
traceroute to i70pc05.tm.uni-karlsruhe.de (141.3.66.114), 30 hops max, 40 byte packets

 1  corona.ibr.cs.tu-bs.de (134.169.34.1)  0.391 ms  0.150 ms  0.120 ms
 2  ciscobs.rz.tu-bs.de (134.169.246.1)  0.493 ms  0.421 ms  0.799 ms
 3  ar-braunschweig3.g-win.dfn.de (188.1.46.137)  0.683 ms  0.598 ms  0.724 ms
 4  cr-hannover1-po3-0.g-win.dfn.de (188.1.88.65)  2.962 ms  3.375 ms  2.116 ms
 5  cr-frankfurt1-po9-3.g-win.dfn.de (188.1.18.181)  10.483 ms  9.987 ms  10.179 ms
 6  cr-stuttgart1-po4-0.g-win.dfn.de (188.1.18.78)  11.513 ms  11.454 ms  11.331 ms
 7  ar-stuttgart2-ge0-0-0.g-win.dfn.de (188.1.76.4)  11.826 ms  11.624 ms  11.530 ms
 8  Stuttgart2.belwue.de (188.1.38.54)  54.890 ms  10.080 ms  10.146 ms
 9  Karlsruhe1.belwue.de (129.143.1.4)  12.692 ms  11.383 ms  11.101 ms
10  BelWue-GW.Uni-Karlsruhe.de (129.143.166.130)  11.427 ms  11.153 ms  11.122 ms
11  r-rtr-ospf-1-rnz-164-22a.rz.uni-karlsruhe.de (129.13.191.56)  13.071 ms  12.924 ms  13.148 ms
12  * * *
13  * * *
14  i72tmrt01.tm.uni-karlsruhe.de (141.3.70.1)  14.663 ms  13.704 ms  14.409 ms
15  i70pc05.tm.uni-karlsruhe.de (141.3.66.114)  13.997 ms  13.778 ms  13.681 ms
```

(b) From Braunschweig to Karlsruhe

Figure 7.15: Transmission delay between Karlsruhe and Braunschweig acquired by "passive" Traceroute program

node system more easily and to make the test more directly. Four applications are initiated at three different source nodes, and they are executed and therefore consume resources in the same destination node. I.e., a Traceroute application (Traceroute1) is initiated at source node 1, which sends 90 TracerouteCapsules at an interval of 2s to the destination node. After the Traceroute 1 begins, another Traceroute application (Traceroute 2) and an Image application (Image 1) are started at source node 2 one after another. Traceroute 2 sends TracerouteCapsules to the same destination node at an interval of 1.5s, and the Image 1 requests an image from the destination node. And at the same time, another image application (Image 2) is started at source node 3; it requests also an image from the same destination node. The source node 2 and 3 are located in Karlsruhe, and the source node 1 and the destination node are in Braunschweig.

The desired resource requirement r_d and the corresponding ARVS of the four appli-

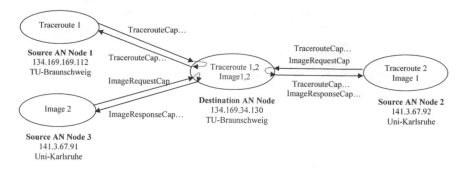

Figure 7.16: Network configuration and applications for resource utilization tests

cations are summarized in Figure 7.17.

		Traceroute 1	Traceroute 2	Image 1	Image 2
Rd		(300000,300000,400000)	(200000,30000,300000)	(144010,30000,400000)	(36240,20000,60000)
ARVS	RV$_L$	(100000,20000,300000)	(100000,2000,200000)	(100000,25000,100000)	(20000,15000,50000)
	RV$_H$	(500000,4000000,5000000)	(400000,400000,5000000)	(350000,400000,2000000)	(400000,400000,600000)
	Ef	$0.3b+0.3c+0.4m$	$0.2b^2+0.3c^2+0.5m^2$	$0.28b+0.3c+0.14m$	$0.3b^2+0.2c^2+0.6m^2$
	WE	0.1	0.05	0.01	0.05
	HC	0.1	0.05	0.01	0.05

Figure 7.17: Desired resource requirement and adaptation capability of the applications

7.2.2.2 Test Results

Figure 7.18 illustrates the system resource status, the resources acquired by each application, and the system resource utilization in the destination node. Here the system resource utilization is defined as the ratio of the resource already consumed in the system to the total system resource capability. The available resources in the system is recorded at the moment after the admission control exerted on the corresponding application, i.e., the application is either accepted or rejected. The initial value means the available resources in the system before the execution of all these applications.

In this test, before the applications began, the available resources at the destination node were (450000, 69436748, 408719360), and the first application, i.e., Traceroute 1, obtained its desired resources. Thus, before the second application Traceroute 2 came, the available resources in the system were (150000, 69136748, 408319360). As Traceroute 2 arrived, there was not enough bandwidth in the system; thus the resource adaptation within the application began first and it was successful; therefore Traceroute 2 got the resource (126712, 304628, 1305441). After Traceroute 2 was accepted by the system,

		Traceroute 1	Traceroute 2	Image 1	Image 2
r_d		(300000,300000, 400000)	(200000,30000,300000)	(144010,30000,400000)	(36240,20000,60000)
Acquired Resource r_{ac}		(300000,300000,400000)/ (151081, 41428,1290382)	(126712,304628,1305441)	(144010,30000,400000)	(25139,30671,141731)
Any adaptation?		no	within in app	among app	within app
Accepted by the sys.?		yes	yes	yes	yes
Avail. res. in sys. initially (450000, 69436748, 408719360)		(150000, 69136748, 408319360)	(23288, 68832120, 407013919)	(28197, 69060692, 405723537)	(3058, 69030021, 405581806)
Sys. res. utilization (%)	bandwidth	67	94.8	93.7	99.3
	CPU	0.43	0.87	0.54	0.59
	memory	0.1	0.42	0.74	0.77

Figure 7.18: Acquired resources and system resource utilization

the available resources in the system became (23288, 68832120, 407013919). When the third application, i.e., Image 1, started, there was still not enough bandwidth for it, and adaptations began. However, the adaptation within this application could not succeed. Finally, the resources allocated to Traceroute 1 were adjusted to (151081, 41428, 1290382), whereas Image 1 got the desired resources of (144010, 30000, 400000). At this moment, the available resources in the system were (28197, 69060692, 405723537). Again, there was not enough bandwidth in the system for the fourth application Image 2 as it arrived. Nevertheless the adaptation within this application succeeded, and it got the resource (25139, 30671 141731). At the moment when all the four applications ran in the system, the available resources in the system were (3058, 69030021, 405581806).

From this example we can see that the resource utilization degree of the active node system can be improved through the resource adaptation mechanism. If there were no resource adaptation mechanism in the system, the application Traceroute 2 and Image 1 could not be accepted due to the scarcity of the bandwidth resource, although instead Image 2 can be accepted by the system. In this case, the bandwidth resource utilization in the system is 74.72%. I.e., when the bandwidth resource utilization in the system is only 74.72%, already 2 applications have been rejected. However, through the resource adaptation mechanism, the resources can be re-organized and re-distributed among applications, and all the 4 applications can be accepted by the system. Finally, the bandwidth utilization can reach 99.3%.

7.2.3 Effects of Adaptation

The adaptive resource admission control mechanism may cause some changes to the applications in case resource adaptations related to the applications occur. In this section, we illustrate a possible effect of the adaptive admission control mechanism on the performance of the applications quantitatively through an example, although theoretically there is no direct relationship between the performance of an application and the adaptive admission control mechanism in the AN node system. And thereby we also show how the adaptive feature of the node system can be utilized by the applications.

Hence, in the following test, we limit the available resources in the system and force adaptations to happen during the admission control, to observe the performance of the same application when executed under different resource conditions.

7.2.3.1   Test Network

The network configuration of the test is shown in Figure 7.19. Here, three active nodes have been used. Two of them are located in Karlsruhe. They both act as clients, and each of them can initiate multiple Image applications requesting images from the server. The other node is in Braunschweig and acts as the image server. This time we let only different Image applications run in the test network and constrain the available resources in the server node system, in order to observe the resources that each application acquires and the corresponding image quality.

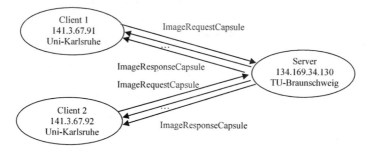

Figure 7.19: Network configuration for effects on performance tests

7.2.3.2   Tests and Results

In the following test, all the Image applications have the same desired resource require-ment r_d and the adaptability ARVS, namely, r_d=(200000, 300000, 600000), RV_L=(20000, 30000, 60000), RV_H=(300000, 4000000, 7000000), and the efficacy function Ef=$0.1b^2$+$0.1m^2$ +$0.2c^2$, WE=HC=0.05. All the applications request an image with the original size of 512×384 pixels from the same image server. Before the first application started, the avail-able resources in the server node system were (250000, 70000000, 336150528), i.e., the network bandwidth was restricted. After Client 1 initiated two different image requests to the server one after another, Client 2 sent immediately also two image requests to the server one after another. Therefore there were altogether 4 Image applications in the server node for some time. The resources that each Image application acquired and the corresponding image quality, described by the image size, format and compression level are summarized in Figure 7.20. Since the above 4 Image applications are completely identical, to some extent, it can also be considered as one application executed under different resource conditions.

As the first image request from Client 1 arrived, there were enough resources in the server node system and the application got its desired resources. Thus, the user obtained the original image he wanted, both for the image size and format. But when the second request arrived, the desired resources were not available, and an adaptation within the application occurred. According to the resources that this application acquired, a JPEG image with the compression quality of 80 was sent to Client 1. Similarly, when the third and fourth image request from the Client 2 arrived, according to the status of the available resources in the server system, the applications acquired different amount of resources. And therefore, the user at Client 2 acquired images with different image size and compression level.

Cli-ent	App No.	Acquired Resources	Transferred Data Length	Data Seg. No.	Processing Time (ms)	Total Delay (ms)	Image Quality		
							Size	Format	Compression level
1	1	(200000,300000, 600000)	527622	103	30.03	3066	512×384	.png	No
	2	(76428,450218,6013782)	62649	13	1532.2	2553	512×384	.jpg	80
2	3	(51200,431920,6237891)	34148	7	1652.4	2433	512×384	.jpg	50
	4	(25132,2024831,3078901)	19392	4	160.2	952	256×192	.jpg	80

Figure 7.20: Resource adaptation and image quality

In this test, the maximal capsule length is set to 5114 bytes. That means the image data has to be segmented and sent in several capsules to the users. Other data transmitted in each capsule include the image file name, length, the number of the total segments, the IP address of the server, the data segment number, and the compression quality if the image has been compressed. Together with the header of the capsule, the maximal image data length in each ImageResponseCapsule is 5000 bytes. In Figure 7.20, the processing time denotes the interval between the time when the ImageRequestCapsule is received and when the first ImageResponseCapsule is sent at the server node. It includes the time for resource admission control and adaptation, as well as for image processing such as compressing and changing the image size. The total delay means the interval between the time when the ImageRequestCapsule is sent at the clients and when all the ImageResponseCapsules are received.

Here we can see that our node system tries to accept each application request as far as possible through the resource adaptation technique. When the resource adaptability of an application is provided to the network node, the node system can adjust its execution according to the system resource status.

7.3   Conclusion and Discussion

In this chapter, we have evaluated the ANwithARM node system through running some active applications in the test network consisting of such active nodes. We have examined the processing overhead introduced by the ANwithARM system, and analyzed the improvement of the system resource utilization achieved by the adaptation mechanism. We have also demonstrated the effects of the adaptation mechanism on the performance of the applications and shown how applications can take advantage of the adaptive feature provided by the node system.

The results show that the adaptive admission control mechanism in ANwithARM can improve the resource utilization of the whole system through balancing the resources of different types and redistributing resources inside and among applications in the node system. Therefore, the possibility that an application is accepted by the AN nodes is increased. We believe that the introduced overhead is acceptable.

However, since we concentrate on the adaptive admission control mechanism of the node system, and evaluate if our mechanism can really improve the total resource utilization of the node system, we have not paid much attention to the active applications implemented on our active node architecture, e.g., if these applications are practical, if an image server should store multiple image file versions or using the transcoding method to provide images files with different image size, format and compression levels etc. Since this is not the goal of our work.

Moreover, since the network bandwidth resource in the node system is relative easy to be controlled, we have controlled the occurrence of resource adaptations only through limiting the available bandwidth resource in the system. In addition, the numbers of applications running simultaneously in one active node is small, we have not performed scalability test. Besides that, the test network consists of only three or four active nodes, and is quite small.

Chapter 8

Summary and Conclusions

This dissertation has examined the resource usage in the AN nodes and studied the resource adaptability from the perspective of the relationship between different types of resources. It has suggested a method for describing the resource usage and the resource adaptability of applications in ANs. An adaptive admission control mechanism for the AN nodes has been proposed and an AN node architecture supporting this mechanism has been implemented. This chapter summarizes the work in this dissertation, draws conclusions and makes some suggestions for future work.

8.1 Summary

Chapter 1 discussed the problems exerted on resource usage which are caused by the introduction of ANs. There, we analyzed the general characteristics related to resource usage in AN nodes and presented the thesis of this dissertation. Namely a new approach for the provisioning of resource guarantees for applications in ANs should be proposed. Furthermore, new problems introduced by the use of multiple types of resources, such as the complementarity of different types of resources, the flexibility of applications related to the resource usage as well as the unbalanced use of system resources, must be taken into account in order to make the system resources well utilized.

Chapter 2 introduced the background information related to the research work in this dissertation. It discussed the main features of the ANs through comparing ANs with the traditional IP-based networks, and gave an overview of the key issues involved by the active networking technology, including the basic approaches to realize the programmability, the general architecture of an AN node, the resource organization in the node system, the active network encapsulation protocol, as well as the methods for applications to realize the demanded functions in ANs. Moreover, the related technologies for making networks programmable, namely the open programmable interface networks and mobile agents, were also reviewed. In addition, two general resource management methods used in the traditional IP-based networks for providing QoS for applications, i.e., resource reservation and adaptation, were discussed. Related work about the AN node

175

architecture and the resource management in the active and programmable networks was also surveyed.

Chapter 3 examined the resource usage characteristics in ANs in detail and suggested the resource vector (RV) concept as the method to describe the resource usage in ANs in order to emphasize these characteristics and solve the new emerged problems caused by these characteristics. As a unit, a resource vector can represent all the resources needed by an application. As an individual, each resource type and amount needed by the application can be stressed. By using RVs, both the amount and the type of the resources needed and used by applications can be represented. Furthermore, two categories of operations are introduced for RVs, including direction difference, compare, division etc. especially designed for RVs. Through these operations, various resource states in the node system can be calculated. In general, RV provides the foundation for the whole work in this dissertation. Based on the concept and operations of RVs, the total system resources and the resources consumed by all the applications in the system can be expressed by RVs in the resource space. This has led to the resource adaptation method introduced in chapter 5, which assigns the system resources to various applications while keeping the usage of different types of resources in the system in balance. RV provides also a basis for using ARVS to describe the adaptability of applications as introduced in chapter 4, which takes the flexibility of applications related to resource usage into account.

Chapter 4 studied the adaptability of applications related to resource usage with the purpose of realizing system-side adaptation. Based on the analysis of applications, it is argued in this dissertation that the amount of resources that can satisfy the performance requirements of applications is not unique. Thus, the possible resource consumption of an application can be expressed using multiple RVs. In other words, multiple points in the resource space can be accepted by the applications and provide similar performance. These points construct a closed space, called the adaptable resource vector space (ARVS). The ARVS is defined using the efficacy function, the cost function and resource limits. These concepts represent both the properties of applications themselves and the resource status in the active node system. Therefore, they can be used both to describe the adaptability of an application and to stipulate the resource adaptation scope for the system. Hence, in case the suggested resource requirement of an application cannot be satisfied by the node system, it is possible to select another point in the ARVS according to the resource status in the node system, so that the needs of applications can be satisfied by making full use of the system resource.

In chapter 5, an adaptive admission control mechanism in AN nodes was presented. The most notable characteristic of this mechanism is that it takes both the adaptability of applications and the system resource balance into account during the admission control procedure. In case the resource requirement of an application cannot be satisfied, the application will not be rejected immediately as it is done by other general resource admission control systems. Instead, an adaptation procedure occurs, searching another point in the ARVSs of applications. Three issues are considered during the search in an ARVS: the application performance requirement, the system resource status and the system resource balance.

The resource adaptations are performed first in the ARVS of the new arriving application. In case this first step cannot succeed, an adaptation is done in the ARVS of other applications running in the system. As a result, the system resources can be adjusted among the different types of resources and redistributed among different applications.

Optimization techniques are applied to find a suitable RV in an ARVS to replace the desired RV. The resource requirement of applications, the system resource status and balance, as well as the corresponding ARVSs are decomposed as the objective functions and constraints of the optimization problem. The Simplex optimization algorithm has been selected due to its simplicity and since it does not need Gradient and Hessian of the objective functions. A comparison between the Simplex and the Sequential Quadratic Programming (SQP) optimization algorithm where both have been applied in the resource adaptation procedure was also given. This showed that the Simplex algorithm provides good optimization results within this usage area.

Chapter 6 described the implementation of the active node architecture supporting the suggested adaptive resource admission control mechanism. The node architecture provides a general AN environment through supporting the general EE (Execution Environment) and NodeOS (Node Operating System) functions and the NodeOS APIs suggested by the AN research group. In addition, a resource management subsystem consisting of three modules is integrated in the NodeOS, carrying out the suggested adaptive resource admission control algorithm, schedulers and controls the resources consumed by each application according to the agreed value, and provides resource query functions through maintaining the system resource status. Moreover, in order to configure the node system and observe the diverse resource status in the node system dynamically, a graphical user-node interface and a resource monitor tool were also implemented as an embedded AA (active application) part in the node architecture.

And finally in chapter 7, a resource monitoring tool and two test applications, namely Traceroute and Image, were introduced. All of them were implemented as active applications on top of the realized AN node architecture. The resource monitoring tool can be used to monitor the various resource states in the node system. The Traceroute and the Image application were designed to measure the performance of the implemented node system and illustrate how to use the adaptive admission control mechanism provided by the node system. Experiments using these applications with different scenarios were made to validate the suggested adaptive admission control algorithm in the AN node architecture. And the performance of the whole system, including the introduced overhead, the improvement of the resource utilization and the affects of adaptations on applications, were also examined on the test ANs consisting of the implemented node systems established on the practical Internet. The results illustrate that the suggested adaptive resource management mechanism is feasible, fulfills its purpose and offers suitable performance.

8.2 Contributions

In general, this dissertation has studied the resource management in active nodes from the perspective of the relationship between different types of resources. It argues that multiple types of resources are needed in active nodes and the consumption of different kinds of resources in active nodes is not always balanced. Moreover, due to the use of multiple types of resources, the resource requirement of applications may be not fixed, this provides more adaptation possibility for AN nodes than in the tradition IP-based networks. In conclusion, the contributions of this dissertation lie in the following aspects:

- It has suggested RV to describe the resource usage in ANs, which has well emphasized the characteristics of resource usage in ANs, such as the multi-dimensionality, complementarity, and high-level sharing. RV provides also a foundation for the work in this dissertation.

- It has proposed ARVS to express the adaptability of applications. This has considered the flexibility of applications related to the resource usage, and solved the corresponding problems for realizing the application-specific system-side adaptation.

- It has presented an adaptive resource management mechanism, which has taken both the provisioning of resource guarantees for applications, and the resource balance and utilization in AN nodes into account. Through this mechanism, the system resources can be adjusted among different types, and re-distributed among different applications according to the needs of applications and the resource status in the node system.

- It has implemented an active node architecture which both provides the basic AN node functions and realizes the suggested resource management mechanism through an explicit resource management subsystem.

- It has evaluated the implemented AN node architecture in the Internet. Several test applications have been developed and run on the test AN consisting of several AN nodes executing the implemented node system. The results show that the suggested adaptive resource management mechanism can be realized with sufficient performance while also providing additional functionality.

8.3 Future Work

This section considers some open problems that have not been addressed in this dissertation due to the lack of available time and test tools. Three issues are involved: further tests and improvements related the node architecture, and some possible extensions of the adaptive admission control mechanism for realizing new functions.

8.3.1   Node Architecture

8.3.1.1   Tests

So far, we have only evaluated the implemented AN node architecture in a small scope. The number of both the AN nodes and the applications running in the nodes is not large. Further tests would be beneficial to evaluate the performance and the scalability of the presented adaptive admission control mechanism and the implemented AN node architecture in large scale environments.

8.3.1.2   Improvements

In order to realize and validate the suggested admission control mechanism, some simplifications are made to parts of the components in the AN node architecture so far. For instance, the policy control issue has not been addressed in full depth. Up to now, only some simple policies have been given. In addition, the System and Network Management EE is also a simple prototype in the current implementation. Before the node architecture could be used in the real networks, more administration and management support should be added into this module.

Moreover, so far our node implements only one type of EE. More types of EEs could be supported by the node architecture. In this case, functions like resource allocation among different types of EEs at the system initiation time, information that can be shared among different types of EEs, as well as the communication between different types of EEs, should be extended.

Regarding the resource adaptation algorithm in the node system, so far the adaptations are triggered only by the admission control. In other words, the lack of resources for the new arrival applications results in the occurrence of the adaptations. However, the influence of the end or leaving of the applications in the system on the adaptation mechanism has been omitted, which comprises also the potential resource improvement possibilities for the existing applications. Future work may also cover this issue.

Of course, the most important issue is to design and develop some meaningful active applications on the suggested AN architecture, which is also an important research effort in the field of ANs.

8.3.2   Possible Extensions

This dissertation has suggested an adaptive admission control mechanism and implemented an active node architecture supporting it. To some degree, the node architecture and the admission control mechanism can also be extended for providing other functions.

For example, some new functions and services can be provided through counting the resources in the node architecture with a much finer granularity. So far we provide only the application-level and the node system-level resource accounting. Based on the techniques and some basic functions already implemented in the resource management subsystem, resource accounting at a much finer level can also be provided. E.g., at the service-level, namely resources consumed by various services provided by the node system can also be counted. Thereby, more functions related to the resource status in the system can also be provided to the applications of the network administrators or the normal end-user users. This would be helpful to the management of the whole ANs and for providing some new services, such as resource-based routing, resource-based service provisioning and so on.

In addition, the adaptive admission control mechanism can also be extended. E.g., the criteria for admitting new applications can be changed or extended. Examples are that the resource capability of the next hop devices, or the resource status or capability of the leaf-nodes can also be added as the criteria for admission control besides the current system resource status. Such that functions like "transcoder" or "adaptive filter" etc. can also be provided by the AN nodes.

The adaptive admission control mechanism proposed in the dissertation has been studied for AN nodes. However, the general idea and method can also be applied to other areas, where multiple types of resources are involved or tradeoff among multiple issues exists. E.g., in application-level server/node system, overlay networks and proxies etc.

8.4   Conclusion

It is the thesis of the dissertation that in AN nodes multiple types of resources must be considered and treated both separately and harmoniously. The need of multiple types of resources and the flexibility of applications may result in the unbalanced usage of node system resources. This weakens the AN nodes in providing services for applications. Therefore, measures concerning the relationship among different types of resources must be taken for the provisioning of resource guarantee for applications, so that the system resources in AN nodes can be well utilized and applications can be better served.

For these considerations, a method for describing the resource usages in AN nodes and the adaptability of applications has been presented, emphasizing the relationship among different types of resources. The suggested ARVS both formulates a generic model for network adaptation, and expresses the individual adaptation requirements of applications. Based on these, an adaptive admission control algorithm is proposed to make trade-off among different kinds of resources and different applications on the one side, and allocate system resources to applications as balanced as possible on the other side. In addition, an AN node architecture with an explicit resource management subsystem has been implemented. The architecture supports both the generic network admission control and the flexible application-specific adaptation. The evaluation demonstrated that this approach is feasible and with acceptable overhead.

Appendix A

Options in Active Packets

As introduced in section 2.2.4, the active network encapsulation protocol (ANEP) [ABG+97] has suggested a general format for active packets. Figure A.1 illustrates the format of the Options mentioned in figure 2.2

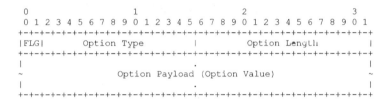

Figure A.1: Options suggested in ANEP

Generally, the options have the form of TLV (Type/Length/Value). The Option Type field identifies the option. How the active node handles the Option Payload depends on the value of the Option Type, i.e., Type ID. The length of this field is 14 bits. The following values have been reserved:

1   Source Identifier

2   Destination Identifier

3   Integrity Checksum

4   N/N Authentication

All values intended for public use are under the authority of the Active Networks Assigned Numbers Authority (ANANA). Other parties can use their own values for this field if the most significant bit (Flags bit 0) is set. In this case, the Options are only meaningful inside the specified evaluation environment, so the proper authority for assigning these values is the Option Type owner.

The Option Length field contains the length of the TLV in 32 bit words. This includes the length of the Flags, Option Type, Option Length and Option Payload fields. This value must never be less than 1 (for an option with a zero sized Option payload). If the Option payload size is larger than the size of the data it carries, it is recommended that the excess 1-3 octets be zero filled and be ignored by a receiving implementation. The length of this field is 16 bits.

The two most significant bits of the first word in an Option (bits 0 and 1) are used as a Flag field. Bit 0 (Private) is used to indicate that the Option Type is only meaningful inside the specified environment. Other nodes should not try to parse the Option at packet receipt if this bit is set. The value of bit 1 of the Flag field defines the action taken if the active node does not know how to process the indicated Option Type.

0   Ignore this option and continue processing the header. It is recommended that the active node logs the event.

1   Discard the packet. It is recommended that the active node logs the event.

Among the four options described in ANEP, the Source and the Destination Identifier option have the format as shown in figure A.2:

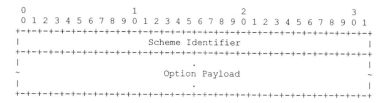

Figure A.2: Source and Destination Identifier Option

Here, the following values have been reserved for the Scheme Identifier:

1   IPv4 address (32 bits)

2   IPv6 address (128 bits)

3   802.3 address (48 bits)

IPv4 and IPv6 addresses are naturally aligned to 32 bits. For 802.3 addresses, the remaining two bytes of the Option payload should be set to zero.

According to the above suggestions, so far we have set the bit 0 of the Flag field to indicate the defined options are for private use. Bit 1 of the Flag field is placed to 0, so that the options are neglected and the packet can be processed continuously in case the option cannot be recognized due to errors. For the moment, we have defined 3 options. Their Type ID are:

11   Resource Vector (RV)

12   Adaptive resource vector space (ARVS)

13   Concerned Node Addresses (ConcernedNode)

In the following sections, we describe the format of these options in detail.

A.1   Resource Vector (RV)

This option includes the values of the bandwidth, CPU and memory resource requested by an application. Each resource type is denoted using a 32 bit value, as shown in figure A.3.

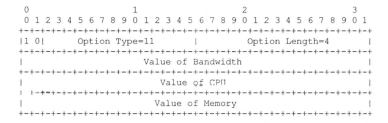

```
 0                   1                   2                   3
 0 1 2 3 4 5 6 7 8 9 0 1 2 3 4 5 6 7 8 9 0 1 2 3 4 5 6 7 8 9 0 1
+-+-+-+-+-+-+-+-+-+-+-+-+-+-+-+-+-+-+-+-+-+-+-+-+-+-+-+-+-+-+-+-+
|1 0|      Option Type=11       |         Option Length=4        |
+-+-+-+-+-+-+-+-+-+-+-+-+-+-+-+-+-+-+-+-+-+-+-+-+-+-+-+-+-+-+-+-+
|                       Value of Bandwidth                       |
+-+-+-+-+-+-+-+-+-+-+-+-+-+-+-+-+-+-+-+-+-+-+-+-+-+-+-+-+-+-+-+-+
|                         Value of CPU                           |
+-+-+-+-+-+-+-+-+-+-+-+-+-+-+-+-+-+-+-+-+-+-+-+-+-+-+-+-+-+-+-+-+
|                       Value of Memory                          |
+-+-+-+-+-+-+-+-+-+-+-+-+-+-+-+-+-+-+-+-+-+-+-+-+-+-+-+-+-+-+-+-+
```

Figure A.3: Resource Vector (RV) Option

A.2   Adaptable Resource Vector Space (ARVS)

This option includes the adaptability information of applications. Figure A.4 illustrates its format.

The first six 32 bits in the payload field are used to denot RV_L and RV_U respectively. For HC and WE, each uses 16 bits with a value as specified in section 5.2.2. Since the length of the efficacy function (see figure 5.3) is variable, the total length of this option depends on the length of the efficacy function. If there are 1-3 octets of excess after the efficacy function, zero is filled and ignored at the receiving node.

A.3   Concerned Node Address

This option is so far used by the Traceroute application, to indicate the addresses of the nodes where information of the node system should be collected and carried back to the source. Figure A.5 illustrates the format of this option.

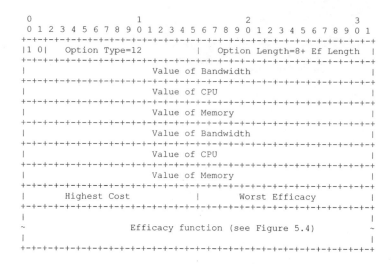

Figure A.4: Adaptable Resource Vector Space (ARVS) Option

Here, the Scheme Identifier field defined for the Source and the Destination Identifier option in ANEP is used. The length of this option depends on the number of the concerned nodes and the address scheme used. For the moment, IPv4 address scheme is adopted in our node system. Therefore, the length is equal to the number of the concerned nodes plus 2.

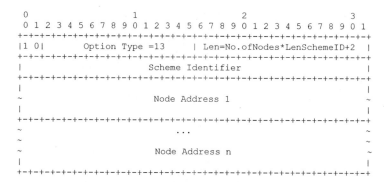

Figure A.5: Concerned Node Address Option

Bibliography

[AAH+98]   D.S. Alexander, W.A. Arbaugh, M.W. Hicks, P. Kakkar, A.D. Keromytis, J.T. Moore, C.A. Gunter, S.M. Nettles and J.M. Smith, "The SwitchWare Active Network Architecture". IEEE Network Magazine, Special Issue on Active and Controllable Networks, 12(3):29-36, May 1998.

[AAK98]   D.S. Alexander, W.A. Arbaugh, A.D. Keromytis, and J.M. Smith, "A Secure Active Network Environment Architecture". IEEE Network Magazine, Special Issue on Active and Controllable Networks, 12(3):37-45, May 1998.

[AAS97]   T.F. Abdelzaher, E.M. Atkins and K.G. Shin, "QoS Negotiation in Real-time Systems and its Application to Automated Flight Control". Real-Time Technology and Applications Symposium, 1997.

[ABF97]   W. Almesberger, J.L. Boudec and T. Ferrari, "Scalable Resource Reservation for the Internet". In Proc. of IEEE Conference on Protocols for Multimedia Systems - Multimedia Networking, Nov. 1997.

[ABG+97]   D.S. Alexander, B. Braden, C.A. Gunter, A.W. Jackson, A.D. Keromytis, G.J. Minden and D. Wetherall, "Active Network Encapsulation Protocol (ANEP)". http://www.cis.upenn.edu/switchware/ANEP/docs/ANEP.txt, July 1997.

[AHI+00]   K.G. Anagnostakis, M.W. Hicks, S. Ioannidis, A.D. Keromytis and J.M. Smith, "Scalable Resource Control in Active Networks". In Proc. of the 2nd International Working Conference on Active Networks (IWAN'00), Tokyo, Japan, Oct. 2000.

[Ale98]   D.S. Alexander, "ALIEN: A Generalized Computing Model of Active Networks". Ph.D Dissertation, Department of Computer and Information Science, University of Pennsylvania, Sept. 1998.

[AN01]   AN Node OS Working Group, "Node OS Interface Specification". Jan. 10, 2001.

[Ants]   ANTS-an Active Node Transfer System. http://www.cs.washington.edu/research/networking/ants/

[ASN+97]   D.S. Alexander, M. Shaw, S.M. Nettles and J.M. Smith, "Active Bridging". In Proc. of the ACM SIGCOMM'97 Conference, pp. 101-111, Sept. 1997.

[BBC+98]   S. Blake, D. Black, M. Carlson, E. Davies, Z. Wang and W. Weiss, "An Architecture for Differentiated Services". RFC 2475, Dec. 1998.

[BBD+97] R. Black, P. Barham, A. Donnelly and N. Stratford, "Protocol Implementation in a Vertically Structured Operating System". In Proc. of the 22nd IEEE Conference on Local Computer Networks (LCN'97),Minnesota, U.S.A, Nov. 1997.

[BBG99] J. Blanquer, J. Bruno, E. Gabber, M. Mcshea, B. zden and A. Silberschatz, "Resource Management for QoS in Eclipse/BSD". In Proc. of the FreeBSD 1999 Conference, Berkeley, California, Oct. 1999.

[BBR00] S. Berson, B. Braden and L. Ricciulli, "Introduction to the ABone". http://www.isi.edu/abone/

[BCEL] The Byte Code Engineering Library. http://jakarta.apache.org/bcel/

[BCL98] G. Bianchi, A.T. Campbell and R.R.-F. Liao, "On Utility-Fair Adaptive Services in Wireless Networks". In Proc. of the IEEE/IFIP International Workshop on Quality of Service (IWQoS'98), Napa Valley, USA, May 1998.

[BCL99] I. Busse, S. Covaci and A. Leichsenring, "Autonomy and Decentralization in Active Networks: A Case Study for Mobile Agents". In Proc. of the 1st International Working Conference on Active Networks (IWAN'99), Berlin Germany, June/July 1999.

[BCS94] R. Braden, D. Clarkand and S. Shenker, "Intergrated Services in the Internet Architecture: an Overview". RFC 1633, June, 1994.

[BCZ96] S. Bhattacharjee, K.L. Calvertand and E.W. Zegura, "On Active Networking and Congestion". Technical report GIT-CC-96/02, College of Computing, Georgia Institute of Technology.

[BCZ98] S. Bhattacharjee, K.L. Calvert and E.W. Zegura, "Congestion Control and Caching in CANEs". In Proc. of the IEEE International Conference on Caching (ICC'98), 1998.

[BDM99] G. Banga, P. Druschel and J. C. Mogul, "Resource containers: A new facility for resource management in server systems". In Proc. of the 3rd Symposium on Operating Systems Design and Implementation (OSDI'99).

[BFI+99] M. Blaze, J. Feigenbaum, J. Ioannidis and A.D. Keromytis, "The Keynote Trust Management System Version 2". Internet RFC 2704, Sept. 1999.

[BGO+98] J. Bruno, E. Gabber, B. Ozden and A. Silberschatz, "The Eclipse Operating System: Providing QoS via Reservation Domains". In Proc. of the USENIX Annual Technical Conference, June 1998.

[BHL00] G. Back, W.C. Hsieh and J. Lepreau, "Processes in KaffeOS: Isolation, resource management and sharing in Java". In Proc. of the 4th Symposium on Operating Systems Design and Implementation, pp. 333-346, USENIX Association, San Diego, CA, Oct. 2000.

[BHV00] W. Binder, J. Hulaas and A. Villazn, "Resource Control in J-SEAL". Technical report, Cahier du CUI No.124. University of Beneva, Oct. 2000. ftp://cui.unige.ch/pub/tios/papers/TR-124-2000.pdf

[Bin99] W. Binder, "J-SEAL2 - A secure high performance mobile agent system". IAT'99 Workshop on Agents in Electronic Commerce, Hongkong, Dec. 1999.

[Bin01] W. Binder, "Designing and Implementing a Secure, Portable and Efficient Mobile Agent Kernel: The J-SEAL2 Approach". Ph.D Dissertation, Department of Natural Sciences and Computer Science, Technical University of Wien, Apr. 2001.

[BLB+02] R. Braden, B. Lindell, S. Berson and T. Faber, "The ASP EE: An Active Network Execution Environment". In Proc. of 2002 DARPA Active Networks Conference and Exposition (DANCE'02), San Francisco, CA, May 2002.

[BLM+98] J. Biswas, A. Lazar, S. Mahjoub, L.-F. Pau, M. Suzuki, S. Torstensson, W. Wang and S. Weinstein, "The IEEE P1520 Standards Initiative for Programmable Network Interfaces". IEEE Communications Magazine, Oct. 1998.

[BLN+94] R. H. Byrd, P. Lu, J. Nocedal and C. Zhu, "A limited memory algorithm for bound constrained optimization". Technical report NAM-08, Department of Electrical Engineering and Computer Science, Northwestern University, revised May 1994.

[BM98] M. Breugst and T. Magedanz, "Mobile Agents - Enabling Technology for Active Intelligent Network Implementation". IEEE Network, 12(3):53-60, Aug. 1998.

[BS91] T. Bihari and K. Schwan, "Dynamic Adaptation of Real-Time Software". ACM Transactions on Computer System, May 1991.

[BYF+98] Y. Bernet, R. Yavatkar, P. Ford, F. Baker, L. Zhang, K. Nichols and M. Speer, "A Framework for Use of RSVP with Diffserv Networks". IETF Draft, Nov. 1998.

[BZB+97] R. Braden, L. Zhang, S. Berson, S. Herzog and S. Jamin; "Resource Reservation Protocol (RSVP) - Version 1 Functional Specification". RFC 2205, Sept. 1997.

[Cal99] K.L. Calvert et al, "Architectural Framework for Active Networks, Version 1.0". Active Network Working Group Draft, July 27, 1999.

[CCH+98] G. Czajkowski, C. Chang, C. Hawblitzel, D. Hu and T.von Eicken, "Resource Management for Extensible Internet Servers". In Proc. of the 8th ACM SIGOPS European Workshop, Sintra, Portugal, Sept. 1998.

[CE98] G. Czajkowski and T. von Eicken. "Jres: A Resource Accounting Interface for Java". In Proc. of the 1998 ACM OOPSLA Conference, Vancouver, BC, Oct. 1998.

[CEA95] A. Campbell, A. Elefteriadis and C. Aurrecoechea, "End-to-End QoS Management of Adaptive Flows". IEEE Symposium of Multimedia Communications and Video Coding, New York, Oct. 1995.

[Cen77] Y. Censor, "Pareto Optimality in Multi-objective Problems". Appl. Math. Optimization, Vol. 4, pp. 41-59, 1977.

[Chang01] F. Chang, "Automatic Adaptation of Tunable Distributed Applications". Ph. D Dissertation, Department of Computer Science, New York University, Jan. 2001.

[Chiba00] Shigeru Chiba, "Load-time Structural Reflection in Java". In Proc. of the 14th European Conference on Object-Oriented Programming (ECOOP'00), pp. 313-336, Sophia Antipolis and Cannes, France, June 2000.

[CKV+99] A.T. Campbell, M.E. Kounavis, D. Villela, J. Vicente, H.De Meer, K. Miki and K. S. Kalaichelvan, "Spawning Networks". IEEE Network, Vol. 13, No. 4, July/Aug. 1999.

[Clear96] S.H. Clearwater, editor, "Market-Based Control, a Paradigm for Distributed Resource Allocation". World Scientific, 1996.

[CP67] Da Cunha, N.O. and E. Polak, "Constrained Minimization Under Vector-valued Criteria in Finite Dimensional Spaces". Journal of Mathematical Analysis and Applications, Vol. 19, pp. 103-124, 1967.

[CVV99] A.T. Campbell, J. Vicente and D.A. Villela, "Virtuosity: Performing Virtual Network Resource Management". In Proc. of the 7th IEEE/IFIP International Workshop on Quality of Service (IWQOS'99), pp. 65-76, London, June 1999.

[Dahm99] M. Dahm, "Bytecode Engineering". In Proc. of Java Information Tage 1999 (JIT'99), Sept. 1999.

[DAML] http://www.daml.org

[DGL+97] T. Dewitt, T. Gross, B. Lowekamp, N. Miller, P. Steenkiste and J. Subhlok, "ReMos: a Resource Monitoring System for Network Aware Applications". Technical report CMU-CS-97-194, School of Computer Science, Carnegie Mellon University, Dec. 1997.

[DGM+01] S. Dawson, F. Gilham, M. Molteni, L. Ricciulli and S. Tsui, "User Guide to Anetd 1.6.9". http://www.isi.edu/abone/.

[DKS89] A. Demers, S. Keshav and S. Shenker, "Analysis and Simulation of a Fair Queuing Algorithm". In Proc. of the ACM SIGCOMM'89 Conference, Sept. 1989.

[DON] ftp://ftp.mathematik.tu-darmstadt.de/pub/department/software/opti/

[DPS02] H. Dandekar, A. Purtell and S. Schwab, "AMP: Experiences in building an exokernel-based platform for active networking". In Proc. of 2002 DARPA Active Networks Conference and Exposition (DANCE'02), San Francisco, CA, May 2002.

[FBB+97] B. Ford, G. Back, F. Benson, J. Lepreau, A. Lin and O. Shivers, "The Flux OSKit: A substrate for OS and language research". In Proc. of the 16th ACM Symposium on Operating Systems Principles, pp. 38-51, St.Malo, France, Oct. 1997.

[FIPA] http://www.fipa.org

[Fle86] R. Fletcher, "Practical Methods of Optimization, 2nd Edition". A Wiley-Interscience Publication. John Wiley & Sons. ISBN 0471-49463-1, May 2000.

[GMC01a] V. Galtier, K. L. Mills, Y. Carlinet, S. Bush and A. Kulkarni, "Predicting and Controlling Resource Usage in a Heterogeneous Active Network". In Proc. of WOAMS 2001, San Francisco, Aug. 6, 2001

[GMC01b] V. Galtier, K. Mills, Y. Carlinet, S. Bush and A. Kulkarni, "Predicting Resource Demand in Heterogeneous Active Networks". http://w3.antd.nist.gov/ active-nets/.

[GSW97] A. Gupta, D.O. Stahl and A.B. Whinston, "Priority pricing of integrated services networks". Internet Economics, edited by Lee McKnight and J.P. Bailey, MIT Press, 1997.

[HCC+98] C. Hawblitzel, C. Chang, G. Czajkowski, D. Hu and T.von. Eicken, "Implementing Multiple Protection Domains in Java". In Proc. of Annual USENIX Conference, New Orleans, LA, June 1998.

[HKM+98] M. Hicks, P. Kakkar, J. Moore, C.A. Gunter and S. Nettles, "PLAN: A Packet Language for Active Networks". In Proc. of the 3rd ACM SIGPLAN International Conference on Functional Programming Languages, 1998.

[HKS02] M. Hicks, A. Keromytis, J.M. Smith, "A Secure PLAN (Extended Version)". In Proc. of 2002 DARPA Active Networks Conference and Exposition (DANCE'02), San Francisco, CA, May, 2002.

[HMA+99] M. Hicks, J. Moore, D. S. Alexander, C. Gunter and S. Nettles, "PLANet: An Active Internetwork". In Proc. of IEEE INFOCOM'99, Mar. 1999.

[HMN01] M. Hicks, J. Moore and S. Nettles, "Compiling PLAN to SNAP". In Proc. of the 3rd International Working Conference on Active Networks (IWAN'01), Pennsylvania, USA, Sept./Oct. 2001.

[HSN+97] D. Hull, A. Shankar, K. Nahrstedt and J.W.S. Liu, "An end to-end QoS model and management architecture". IEEE Workshop on Middleware for Distributed Real-time Systems and Services, Dec. 1997.

[HW96] J. Huang and P.J. Wan, "On Supporting Mission-Critical Multimedia Applications". In Proc. of the 3rd IEEE International Conference on Multimedia Computing and System, June 1996.

[HWD98] J. Huang, P.J. Wan and D.Z. Du, "Criticality- and QoS-Based Multiresource Negotiation and Adaptation". Journal of Real-Time Systems, 1998.

[ITU95] Recommendation ITU-R BT.500-7, "Methodology for the Subjective Assessment of the Quality of Television Pictures". Geneva, Switzerland, Oct. 1995.

[Janos] The Janos Project, http://www.cs.utah.edu/flux/janos/

[JLD+95] M.B. Jones, P.J. Leach, R.P. Draves and J.S. Barrera, "Modular Real-Time Resource Management in the Rialto Operating System". In Proc. of the 5th Workshop on Hot Topics in Operating Systems, May 1995.

[JRR97]   M.B. Jones, D. Rosu and M.C. Rosu, "CPU Reservations and Time Constraints: Efficient, Predictable Scheduling of Independent Activities". In Proc. of the 16th ACM Symposium on Operating Systems Principles. Saint-Malo, France, pp.198-211, Oct. 1997.

[KBC99]   S. Karnouskos, I. Busse and S. Covaci, "Agent Based Security for the Active Network Infrastructure". In Proc. of the 1st International Working Conference on Active Networks (IWAN'99), Berlin, Germany, June/July 1999.

[KCS98]   R. Kravets, K. Calvert and K. Schwan, "Payoff Adaptation of Communication for Distributed Interactive Applications". Journal on High Speed Networking, Special Issue on Multimedia Communications, 1998.

[KH98]   R. Keller and U. Hlzle, "Binary Component Adaptation". In Porc. of the 12th European Conference on Object-Oriented Programming (ECOOP'98), Brussels, Belgium, July 1998.

[Khan98]   Md. S. Khan, "Quality Adaptation in a Multisession Multimedia System: Model, Algorithms and Architecture". Ph.D. Dissertation, Department of ECE, University of Victoria, 1998.

[KL97]   S. Khan and K.F. Li, "The Utility Model for Adaptive Multimedia Systems". In Proc. of the International Conference on Multimedia Modelling, Singapore, Nov.1997.

[KMH+98]   A.B. Kulkarni, G.J. Minden, R. Hill, Y. Wijata, A. Gopinath, S. Sheth, F. Wahab, H. Pindi and A. Nagarajan, "Implementation of a Prototype Active Network". In Proc. of the 1st IEEE Conference on Open Architectures and Network Programming (OPENARCH'98), Apr. 1998.

[KMT98]   F.P. Kelly, A. Maulloo and D. Tan, "Rate Control for Communication Networks: Shadow Prices, Proportional Fairness and Stability". Journal of Operations Research Society, 49(3):237-252, Mar. 1998.

[KRG+02]   R. Keller, L. Ruf, A. Guindehi and B. Plattner, "PromethOS: A Dynamically Extensible Router Architecture Supporting Explicit Routing". In Proc. of the 4th International Working Conference on Active Networks (IWAN'02), Zurich, Switzerland, Dec. 2002.

[LBL95]   A.A. Lazar, S.K. Bhonsle and K.S. Lim, "A Binding Architecture for Multimedia Networks". Journal of Parallel and Distributed Computing, 30(2):204-216, Nov. 1995.

[LC01]   R.R.-F. Liao and A.T. Campbell, "A Utility-Base Approach for Quantitative Adaptation in Wireless Packet Networks". ACM Journal on Wireless Networks (WINET), Vol.7, No.5, pp. 541-557, Sept.2001.

[Lep01]   J. Lepreau, "Janos Project: FY2001". June 5, 2001. http://www.cs.utah.edu/flux/talks/janos-pim-0601.htm.

[Ler97]   X. Leroy, "Objective Caml". INRIA, 1997. http://caml.inria.fr/ocaml.

[LL73]   C. Liu and J. Layland, "Scheduling Algorithms for Multiprogramming in a Hard Real-time Environment". Journal of the ACM, 20(1):46-61, Feb. 1973.

[LL99]   S.H. Low and D.E. Lapsley, "Optimization Flow Control, I: Basic Algorithm and Convergence". IEEE/ACM Transactions on Networking, 7(6):861-75, Dec. 1999.

[LLM96]   A.A. Lazar, K.S. Lim and F. Marconcini, "Realizing a Foundation for Programmability of ATM Networks with the Binding Architecture". IEEE Journal on Selected Areas in Communications, Special Issue on Distributed Multimedia Systems, Vol. 14, No.7, pp. 1214-1247, Sept. 1996.

[LMB+96]   I.M. Leslie, D.Mc. Auley, R. Black, T. Roscoe, P. Barham, D. Evers, R. Fairbairns and E. Hyden, "The Design and Implementation of an Operating System to Support Distributed Multimedia Applications". IEEE Journal on Selected Areas in Communications, Sept. 1996.

[LWX03]   Y. Li, L. Wolf and F. Xu, "A Remote Resource Monitoring Approach in Active Networks". In Proc. of Workshop on End-to-End Monitoring Techniques and Services (E2EMON) at the 6th IFIP/IEEE International Conference on Management of Multimedia Networks and Services (MMNS 2003), Belfast, Northern Ireland, Sept. 2003.

[LY99]   T. Lindholm and F. Yellin, "The Java Virtual Machine Specification, second edition". http://java.sun.com/docs/books/vmspec/2nd-edition/html/overview.doc.html.

[LZ97]   H.B. Lee and B.G. Zorn, "BIT: A tool for Instrumenting Java Bytecodes". In Proc. of the USENIX Symposium on Internet Technologies and Systems (ITS-97), Berkeley, Dec. 1997.

[MBB+98]   D. Milojicic, M. Breugst, I. Busse, J. Campbell, S. Covaci, B. Friedman, K. Kosaka, D. Lange, K. Ono, M. Oshima, C. Tham, S. Virdhagriswaran and J. White, "MASIF: The OMG mobile agent system interoperability facility". In Proc. of the Second International Workshop on Mobile Agents, pp. 50–67, Stuttgart, Germany, Sept. 1998.

[MBZ+00]   S. Merugu, S. Bhattacharjee, E. Zegura and K. Calvert, "BOWMAN: A Node OS for Active Networks". In Proc. of IEEE INFOCOM'00, Mar. 2000.

[Men99]   P. Menage, "RCANE: A resource controlled framework for active network services". In Proc. of the 1st International Working Conference on Active Networks (IWAN'99), Berlin, Germany, June/July, 1999.

[MHN01]   J.T. Moore, M. Hicks and S. Nettles, "Practical Programmable Packets". In Proc. of IEEE INFOCOM'01, pp. 41-50, Apr. 2001.

[Micro1]   http://www.microsoft.com/msdownload/platformsdk/sdkupdate/

[Micro2]   Microsoft, "Performance Monitoring in Windows Platform". http://msdn.Microsoft.com/library/default.asp?url=/library/en-us/perfmon/base/enumprocesses.asp

[Moo02]   J. Moore, "Practical Active Packets". Ph.D Dissertation, Department of Computer and Information Science, University of Pennsylvania, 2002.

[MTH+97] R. Milner, M. Tofte, R. Harper and D.B. MacQueen, "The Standard ML Programming Language (Revised)". MIT Press, 1997.

[MU01]   S. Moyer and A. Umar, "The Impact of Network Convergence on Telecommunications Software". IEEE Communications Magazine, Jan. 2001.

[NBB+98] K. Nichols, S. Blake, F. Baker and D. Black, "Definition of the Differentiated Services Field (DS Field) in the Ipv4 and Ipv6 Headers". RFC 2474, Dec. 1998.

[Nem]   http://www.cl.cam.ac.uk/Research/SRG/netos/nemesis/

[NEOS]   http://www-fp.mcs.anl.gov/otc/Guide/OptWeb/index.html.

[NGK99]   E.L. Nygren, S.J. Garland and M.F. Kaashoek, "PAN: A High-Performance Active Network Node Supporting Multiple Mobile Code Systems". In Proc. of the 2nd IEEE Conference on Open Architectures and Network Programming (OPENARCH'99), Mar. 1999.

[NL00]   K. Najafi and A. Leon-Garcia, "A Novel Cost Model for Active Networks". In Proc. of International Conference on Communication Technologies, World Computer Congress 2000.

[NM65]   J.A. Nelder and R. Mead, "A simplex method for function minimization". Computer Journal 7, pp. 308-313 (1965).

[NS95]   K. Nahrstedt and R. Steinmetz. "Resource Management in Networked Multimedia System". Computer, May 1995.

[NY97]   Q. Ni and Y. Yuan, "A subspace limited memory quasi-Newton algorithm for large-scale nonlinear bound constrained optimization". Mathematics of Computation, Vol.66, No.220, Oct. 1997, pp. 1509-1520, S 0025-5718(97)00866-1.

[OpenSig] Open Signalling Group. http://www.comet.columbia.edu/opensig/

[PGH+01] L. Peterson, Y. Gottlieb, M. Hibler, P. Tullmann, J. Lepreau, S. Schwab, H. Dandekar, A. Purtell, J. Hartman. "An OS Interface for Active Routers". IEEE Journal on Selected Areas of Communication, Vol.19, No. 3, Mar. 2001.

[OR98]   A. Ortega and K. Ramchandran, "Rate-distortion Methods for Image and Video Compression". IEEE Signal Processing Magazine, 15(6):23-50, Nov. 1998.

[Pos81]   J. Postel, "Internet Protocol". RFC 791, Sept. 1981.

[PS98]   P.P. Pan and H. Schulzrinne, "YESSIR: A simple reservation mechanism for the Internet". In Proc. of International Workshop on Network and Operating System Support for Digital Audio and Video (NOSSDAV), July 1998.

[RJM+98] R. Rajkumar, K. Juvva, A. Molano and S. Oikawa, "Resource Kernels: A Resource-Centric Approach to Real-Time Systems". In Proc. of SPIE/ACM Conference on Multimedia Computing and Networking, Jan., 1998.

[RLL+97]   R. Rajkumar, C. Lee, J. Lehoczky and D. Siewiorek, "A Resource Allocation Model for QoS Management". In Proc. of 18th IEEE Real-Time System Symposium, San Francisco, CA, Dec. 1997.

[RS96]   D.I. Rosu and K. Schwan. "Improving Protocol Performance by Dynamic Control of Communication Resources". In Proc. of the 2nd IEEE International Conference on Engineering Complex Computer Systems, 1996.

[RS99]   D. Raz and Y. Shavitt, "An Active Network Approach to Efficient Network Management". Technical report DIMACS 99-25, Center for Discrete Mathematics and Theoretical Computer Science.

[RSY98]   D.I. Rosu and K. Schwan, S. Yalamanchili, "FARA: A Framework for Adaptive Resource Allocation in Complex Real-Time Systems". In Proc. of IEEE Real-Time Technology and Applications Symposium, June 1998.

[RW02]   A. Rudys and D.S. Wallach, "Enforcing Java Run-Time Properties Using Bytecode Rewriting". In Proc. of International Symposium on Software Security, Tokyo, Japan, Nov. 2002.

[SDB+02]   D. Sterne, K. Djahandari, R. Balupari, W.L. Cholter, B. Babson, D. Wilson, P. Narashiman and A. Purtell, "Active Network based DDoD Defense". In Proc. of 2002 DARPA Active Networks Conference and Exposition (DANCE'02), San Francisco, CA, May 2002.

[SDK]   Microsoft Platform Software Development Kit. http://www.microsoft.com/msdownload/platformsdk/sdkupdate/.

[SDL99]   M. Shankar, M. DeMiguel and J. Liu, "An end-to-end QoS management architecture". In Proc. of Real-Time Applications Symposium, June 1999.

[SGW+02]   N. Shalaby, Y. Gottlieb, M. Wawrzoniak and L. Peterson, "Snow on Silk: A NodeOS in the Linux Kernel". In Proc. of the 4th International Working Conference on Active Networks (IWAN'02), Zurich, Switzerland, Dec. 2002.

[Shan74]   C.E. Shannon, "Coding Theorems for a Discrete Source with a Fidelity Criterion". IRE Nat. Conv. Rec.,4:142-163, 1959. Reprinted in D. Slepian (ed.), Key Papers in the Development of Information Theory, IEEE Press, 1974.

[Shen95]   S. Shenker, "Fundamental Design Issues for the Future Internet". IEEE Journal on Selected Areas in Communications, 13(7):1176-1188, Sept. 1995.

[SHH62]   W. Spendley, G.R. Hext and F.R. Himsworth, "Sequential Application of Simplex Designs in Optimization and Evolutionary Operation". Technometrics 4, pp. 441- 461 (1962).

[SJ03]   F. Sabrina and S. Jha, "An Adaptive Resource Management Architecture for Active Networks". Telecommunication Systems, 24:2-4, pp.139-166, 2003.

[SJS99]   B. Schwartz, A. Jackson, T. Strayer, W. Zhou, D. Rockwell and C. Partridge. "Smart Packets for Active Networks". In Proc. of the 2nd IEEE Conference on Open Architectures and Network Programming (OPENARCH'99), Mar. 1999.

[SKB+01] M. Sanders, M. Keaton, S. Bhattacharjee, K, Calvert, S. Zabale and E. Zegura, "Active Reliable Multicast on CANEs: A Case Study". In Proc. of the 4th IEEE Conference on Open Architectures and Network Programming (OPE-NARCH'01), Alaska, USA, Apr. 2001.

[SM99] N. Stratford and R. Mortier, "An Economic Approach to Adaptive Resource Management". In Proc. of HotOS-VII, Mar. 1999.

[SMY+99] K. Sugauchi, S. Miyazaki, K. Yoshida, K. Nakane, S. Covaci and T. Zhang, "Flexible Network Management Using Active Network Framework". In Proc. of the 1st International Working Conference on Active Networks (IWAN'99), Berlin Germany, June/July 1999.

[Spe98a] P. Spellucci, "An SQP Method for General Nonlinear Programs Using Only Equality Constrained Subprobmes". Math. Prog. 82, (1998), pp. 413-448.

[Spe98b] P. Spellucci, "A New Technique for Inconsistent Problems in the SQP Method". Math. Meth. of Oper. Res. 47, (1998), pp. 355-400.

[SRC84] J.H. Saltzer, D.P. Reed and D.D. Clark, "End-to-End Arguments in System Design". ACM Transactions on Computer Systems, Nov. 1984.

[Sun] Sun Microsystems, "JavaTM Virtual Machine Profiler Interface (JVMPI)". http://java.sun.com/j2se/1.3/docs/guide/jvmpi/jvmpi.html

[TGS+96] D.L. Tennenhouse, S.J. Garland, L. Shrira and M.F. Kaashoek, "From Internet to ActiveNet". Request for Comments, Jan. 1996.

[THL01] P. Tullmann, M. Hibler and J. Lepreau. "Janos: A Java-oriented OS for Active Network Nodes". IEEE Journal on Selected Areas in Communications, 19(3), Mar. 2001.

[Tschu93] C. Tschudin, "On the Structuring of Computer Communications". Ph.D Dissertation, Department of Informatics, University of Geneva, Switzerland, 1993.

[Tschu93] C. Tschudin, "Open Resource Allocation for Mobile Code". In Proc. of the $1^{s}t$ International Workshop on Mobile Agents (MA'97), Berlin, Apr. 1997.

[Tschu99] C. Tschudin, "An Active Networks Overlay Networks (ANON)". In Proc. of the $1^{s}t$ International Working Conference on Active Networks (IWAN'99), Berlin, Germany, June/July 1999.

[TW96] D. Tennenhouse and D. Wetherall, "Towards an Active Network Architecture". In Proc. of Multimedia Computing and Networking, San Jose, CA, 1996.

[VCV01] D. Villela, A.T. Campbell and J. Vicente, "Virtuosity: Programmable Resource Management for Spawning Networks". Computer Networks, Special Issues on Active Networks, 2001.

[Wet99a] D. Wetherall, "Active Network Vision and Reality: Lessons from a Capsule-based System". In Proc. of the 17th ACM Symposium on Operating Systems Principles (SOSP'99), Vol.33 of ACM Operating Systems Review, Dec.1999.

[Wet99b] D. Wetherall, "If bandwidth is the scarce resource, then scheduling bandwidth is sufficient". Personal Communication, Dec. 1999.

[WGT98] D. Wetherall, J. Guttag and D. Tennenhouse, "ANTS: A Toolkit for Building and Dynamically Deploying Network Protocols". In Proc. of the 1st IEEE Conference on Open Architectures and Network Programming (OPENARCH'98), San Francisco, CA, Apr. 1998.

[WHH+92] C.A. Waldspurger, T. Hogg, B.A. Huberman, J.O. Kephart and W.S. Stornetta, "Spawn: A Distributed Computational Economy". IEEE Transactions on Software Engineering, 18(2):103-117, Feb. 1992.

[WJK+02] L. Westberg, M. Jacobsson, G. Karagiannis, S. Oosthoek, D. Partain, V. Rexhepi, R. Szabo and P. Wallentin, "Resource Management in Diffserv (RMD) Framework". IETF Internet-Draft. Feb. 2002.

[WLG98] D. Wetherall, U. Legedza and J. Guttag, "Introducing New Internet Services: Why and How". IEEE Network, May/June 1998.

[WW96] C.A. Waldspurger and W.E. Weihl, "An Object-Oriented Framework for Modular Resource Management". In Proc. of the 5th International Workshop on Object Orientation in Operating Systems, pp. 238-143. IEEE Computer Society Press, Oct.1996.

[YC01] D.K.Y. Yau and X. Chen, "Resource Management in Software-Programmable Router Operating Systems". IEEE Journal on selected Areas in communication, Vol.19, No.3, Mar. 2001.

[YG00] L. Yamamoto and G. Leduc, "Resource Trading Agents for Adaptive Active Network Application". Network and Information Systems Journal, Vol.3, 2000.

[Zadeh63] L.A. Zadeh, "Optimality and Nonscalar-valued Performance Criteria". IEEE Transactions on Automatic Control, Vol. AC-8, pp. 1, 1963.

[ZBH+97] L. Zhang, S. Berson, S. Herzog and S. Jamin, "Resource Reservation Protocol". Version 1 Functional Specification, RFC2205, Sept. 1997.

[ZBL+97] C. Zhu, R.H. Byrd, P. Lu and J. Nocedal, "Algorithm 778: L-BFGS-B: Fortran Subroutines for Large-Scale Bound-Constrained Optimization". ACM Transactions on Mathematical Software, Vol.23, No.4, Dec. 1997, pp. 550-560.

[Zhang90] L. Zhang, "Virtual Clock: A New Traffic Control Algorithm for Packet Switching Networks". In Proc. of the ACM SIGCOMM'90 Conference, Sept. 1990.